The Enrichment of Copper Sulfide Ores

by J.D. Clark

with an introduction by Kerby Jackson

Introduction

It has been over a century since the University of New Mexico released it's important publication "A Chemical Study On The Enrichment of Copper Sulfide Ores". First released in 1913, this work has been unavailable to the mining community since those days, with the exception of expensive original collector's copies and poorly produced digital editions.

It has often been said that "*gold is where you find it*", but even beginning prospectors understand that their chances for finding something of value in the earth or in the streams of the Golden West are dramatically increased by going back to those places where gold and other minerals were once mined by our forerunners. Despite this, much of the contemporary information on local mining history that is currently available is mostly a result of mere local folklore and persistent rumors of major strikes, the details and facts of which, have long been distorted. Long gone are the old timers and with them, the days of first hand knowledge of the mines of the area and how they operated. Also long gone are most of their notes, their assay reports, their mine maps and personal scrapbooks, along with most of the surveys and reports that were performed for them by private and government geologists. Even published books such as this one are often retired to the local landfill or backyard burn pile by the descendents of those old timers and disappear at an alarming rate. Despite the fact that we live in the so-called "Information Age" where information is supposedly only the push of a button on a keyboard away, true insight into mining properties remains illusive and hard to come by, even to those of us who seek out this sort of information as if our lives depend upon it. Without this type of information readily available to the average independent miner, there is little hope that our metal mining industry will ever recover.

This important volume and others like it, are being presented in their entirety again, in the hope that the average prospector will no longer stumble through the overgrown hills and the tailing strewn creeks without being well informed enough to have a chance to succeed at his ventures.

Kerby Jackson
Josephine County, Oregon
October 2014

WHOLE NUMBER 75

| Chemistry Series | Albuquerque, N. M., June, 1914 | Vol. I. No. 2 |

A Chemical Study of the Enrichment of Copper Sulfide Ores

A DISSERTATION

Submitted to the Department of Chemistry and to the
Committee on Graduate Study of the Leland Stanford
Junior University in partial fulfillment of the require-
ments for the degree of Doctor of Philosophy

BY

JOHN DUSTIN CLARK

Preface

Since the time recorded in our earliest history metallic ores have been taken from mines and deposits. Undoubtedly the first discovered deposits were found by accident. The mining of the ore probably followed well defined veins and stopped when such were exhausted. The work of the ignorant prospector of to-day is much like the probable simple procedure of the ancients.

Observations which have led to the science of Geology must have made some miners proficient in predicting the occurrence of mineral bearing veins, etc. It is certain that to-day the mining engineer and mining geologist can, as a result of training and observation, "see beneath the ground", and can successfully direct development work carried on to locate valuable metal-bearing bodies.

The engineer and geologist has made use of many of the sciences in conducting his work. Chemistry has always played an important part, and it seems that its field of usefulness has only been entered upon. It appears that the results of purely theoretical and experimental laboratory work in chemistry may be of the most practical importance when given to these mining experts. This much is certain, that for understanding enrichment in our copper mines much chemical work is necessary, as may be seen when an authority like W. H. Emmons[1] says, (in his publication, "Enrichment of Sulfide Ores",) *"To the chemist this paper is an appeal for more experimental data on the important mineral syntheses involved in the processes."*

DEFINITIONS OF ENRICHMENT.

The meaning of *downward* secondary enrichment and a summary of much of the field observation, laboratory collaboration and commercial significance is given in language readily understood by the average man, yet not lacking in scientific information, by Tolman[1] in his article on "Secondary Sulfide Enrichment of Ores". Here is quoted his brief definition of this *downward* process. "Secondary sulfide enrichment involves three processes: (1) The oxidation of the metallic sulfides; (2) The solution of these chiefly as sulfates, chlorides or bicarbonates; and (3) The precipitation of the metals in solution in the form of secondary sulfides. (a) by the reduction of the sulfates to metallic sulfides by carbonaceous matter, or (b) by precipitation by means of hydrogen sulfide, or (c) by reaction of the metallic salts with unoxidized

---

I—W. H. Emmons, "The enrichment of sulfide ores", Bull. U. S. G. S. No. 529, p. 11.

sulfides below the water level, the latter going into solution as sulfates, (or other salts), and the former precipitating as sulfides. In the broadest sense secondary sulfide enrichment includes deposits of any sulfide precipitated in any way out of the descending meteoric waters which owe their metallic content to the leaching of the overlying rock."

A. F. Rogers[II] has recently applied the term *upward* secondary enrichment to processes by which metal-bearing solutions bring their metals as sulfides from greater, and deposits them at lesser depths, in distinction from processes in which the ore-forming solutions flow *downward*.

It is believed that the results of the present investigation have shown that upward enrichment without oxidation is a very important process. As will be demonstrated, crystalline copper sulfides, whether in the form of chalcopyrite, bornite, covellite or chalcocite, may be brought into such condition as to migrate considerable distances without the intervention of any process of oxidation or any process converting the copper into electrolyte solution. Such agencies, as will be shown, are those producing the colloidal copper sulfides.

I—C. F. Tolman, Jr., Min. and Sci. Press, **106**, p. 38, p. 147, p. 178, (1913).

II—A. F. Rogers, Econ. Geol. **8**, p. 781, (1913).

OXIDATION AND SOLUTION OF COPPER MINERALS.

A review of the literature concerning these steps shows that they may take place in one or more of the following ways:

(1) Oxidation of pyrite to produce sulfuric acid and ferric sulfate, which then attack and dissolve chalcocite or other copper minerals.

(2) Thru the greater electrolytic solution tension of some minerals in contact with others.

(3) Solution of a mineral in pure water.

These three are prominently mentioned and are generally accepted as being effective. To these may be added, as is later shown by the author:

(4) Solution or dispersion of the mineral by carbonated water, and that suggested but hardly proved method:

(5) Solution of the metal as a chloride where pyrite and manganese dioxide have a chloride bearing water percolating thru them.

In addition to the processes which have been suggested in the literature it will be shown in this paper that, (from the chemical point of view at least), the enrichment by the highly dispersed copper sulfides, as such, may be, as a result of this work, looked upon as an important factor in the general process of enrichment.

Going into detail concerning these methods of solution we find that the following equations may represent the oxidation of the pyrite, though as Tolman[I] says, "the exact chemical equations are not known and perhaps will never be written":

1. $FeS_2 + 4\ O = FeSO_4 + S$
2. $FeS_2 + 6\ O = FeSO_4 + SO_2$
3. $FeS_2 + 7\ O + H_2O = FeSO_4 + H_2SO_4$
4. $2\ FeSO_4 + H_2SO_4 + O = Fe_2(SO_4)_3 + H_2O$
5. $2\ Fe_2(SO_4)_3 + 9\ H_2O = 2\ Fe_2O_3 . 3H_2O + 6\ H_2SO_4$
6. $FeS_2 + Fe_2(SO_4)_3 = 3\ FeSO_4 + 2\ S$ and
   $2\ S + 6\ Fe_2SO_4)_3 + 8\ H_2O = 12\ FeSO_4 + 8\ H_2SO_4$

and according to Austin[II] the ferric sulfate puts the copper into solution as:

$Fe_2(SO_4)_3 + Cu_2S = CuSO_4 + 2\ FeSO_4 + CuS$

$Fe_2(SO_4)_3 + CuS + 3\ O + H_2O = CuSO_4 + 2\ FeSO_4 + H_2SO_4$

Buehler and Gottschalk[III] have shown that chalcocite in contact with pyrite goes into solution as a sulfate in the presence of air, because of its greater electrolytic

I—Min. and Sci. Press, Jan. 4, 18, and 25, 1913.

II—Austin, "The Metallurgy of Common Metals" 2d ed., p. 281.

III—Buehler and Gottschalk, "The oxidation of sulfides" Econ. Geol. 5, pp. 28-37; 7, pp. 15-35, 1912.

solution tension, and that pyrite, as would be expected, is little acted upon.

It may be mentioned in passing that this is not the only possible explanation of this phenomenon, which may be explained entirely upon differences of reaction velocity in the two cases. Of course if definite equilibrium potentials were measured there would be no question as to the validity of the explanation but since all copper sulfide minerals show quite variable composition it would be natural to suppose that the electrolytic solution tensions would be extremely variable with different specimens. Until it is possible to prepare chemically pure or nearly chemically pure minerals of the various types any interpretation based upon the study of electrolytic solution tension must be considered ambiguous.

Weigel[I] has shown that metallic sulfides are slightly soluble in pure water without air.

Tolman[II] was of the opinion that copper minerals with an excess of carbon dioxide, in certain cases were carried downward as bicarbonates. The author has made such solutions. Whether the solution is a true electrolytic one of copper bicarbonate or a mere suspension of a copper carbonate is an open question.

I—Oscar Weigel, "Die Loslichkeit von Schwermetalle Sulfide in reinem Wasser", Zeit. phys. Chemie, 58, pp. 293-300, (1907).

II—Suggested in Min. and Sci. Press. Jan. 4, 18, 25, 1913.

Lane[III] believes that the copper in the Michigan mines has been carried downward as chloride solutions and suggests electrolytic migration. That the copper may have been in the chloride form seems wholly tenable judging from the chloride content of the Michigan mine waters.

Lindgren[IV] in summarizing Emmon's work mentions the possibility of solution as chlorides where nascent chlorine has played an important role, this chlorine having been produced in accordance with the question:

$$MnO_2 + 4\ HCl = MnCl_2 + H_2O$$

He gives instances of increased solution of gold when manganese dioxide has been added to a solution containing very dilute hydrochloric acid.

The above five methods of oxidation seem to cover the principles involved. One will find, however, mention of many possibilities of oxidation and solution which combine two or more of the above methods. Considerable work has been done in attempting to discover the order in which different minerals oxidize. No two investigators have succeeded in getting the same results[I]. In all the investigations made the invesigators have failed to consider conditions under which the oxidations took place and their results put the chemist

---

III—A. C. Lane, Mich. Geol. and Biol. Survey. Pub. 6, Geol. Ser. 4, Vol. I, p. 43.
IV—Lindgren, "Mineral Deposits," p. 798, 1913.
I—Lindgren, "Mineral Deposits," p. 787, 1913.

in mind of the old Geoffroy and Bergmann affinity tables.

The essential factors entering into the problem of oxidation are; ore, metal, country rock, fissuring, porosity, climate, water level, rainfall, topography, geological age and history of the deposit. These have to be considered in the examination of any particular mine.

PRECIPITATION OF THE METALLIC SULFIDES.

Various methods have been sugested by which the metals in descending solutions of metallic salts can be precipitated as sulfides. The following tried and untried means seem to cover the possibilities of such precipitations:

(1) Substitution because of solution tension of some mineral precipitant as;

$$Ag_2SO_4 + ZnS = Ag_2S + ZnSO_4$$
$$14\ CuSO_4 + 5\ FeS_2 + 12\ H_2O = 7\ Cu_2\ S +$$
$$5\ FeSO_4 + 12\ H_2SO_4$$

(2) Substitution in the presence of sulfur dioxide or other reducing agent as;

$$4\ CuSO_4 + FeS_2 + SO_2 + 6\ H_2O = 2\ Cu_2S$$
$$+ FeSO_4 + 6\ H_2SO_4.$$

(3) Precipitation by means of hydrogen sulfide

(4) Precipitation by means of free sulfur

(5) Neutralization of descending acidic solutions

(6) Precipitation by means of carbon

(7) Precipitation thru loss of a dispersing agent[I].

This paper has much to do with the last feature.

Under the first heading should be noted the work of Schuermann[II]. He established a series of salts in which the sulfides of any one of the metals thereof will be precipitated at the expense of any sulfide lower in the series. The series is undoubtedly correct for the particular specimens used under the conditions which his experiments were performed. What such an order might be under other conditions one cannot say.

The work of Palmer and Bastin[I] shows that such substitution occurs, the author assumes because of solution tension. They used many minerals and found them to precipitate gold and silver. Two of their equations are here given;

$$2\ NiAs + 5\ Ag_2SO_4 + 3\ H_2O = 2\ NiSO_4 + As_2O_3 + 3\ H_2SO_4 + 10\ Ag$$

$$Cu_2S + 2\ Ag_2SO_4 = 2\ CuSO_4 + Ag_2S + 2\ Ag$$

The first fact to be considered under the heading of substitution in the presence of sulfur dioxide or other reducing agent is that mine waters contain less and less

---

I—Suggestion as to the role of colloids is given by P. Krusch, "Primary and secondary ores considered with especial reference to the gel and the rich heavy metal ores", Min. and Sci. Press, **107**, pp. 418-423, 1913.

II—Ernst Schuermann, "Ueber die Verwandschaft der Schwermetalle zum Schwefel", Liebig's Ann. d. Chem. **249**, p. 326.

I—Palmer and Bastin, Econ. Geol. **8**, No. 2, p. 140, 1913.

oxygen as the depth increases as was shown by Lepsius[II]. Emmons[III] states that the oxygen content and acidity of descending waters diminishes as the depth increases. This is strongly emphasized in his discussion of analyses of mine waters.

Both Tolman[IV] and Winchell[V], (see also experiments No. 9 and No. 10 in this paper), have produced chalcocite films on pyrite from slightly acid solutions of copper sulfate in the presence of sulfur dioxide. Chalcocite was not precipitated in its absence.

W. H. Emmons[VI] suggests that this may be explained by the fact that sulfur dioxide removes any atmospheric oxygen from the solutions. The author will have another conclusion to add to this.

All investigators are firm in the belief that chalcocite is not precipitated in the presence of oxidizing action.

In working up their first paper above mentioned, Buehler and Gottschalk obtained sulfur dioxide where the supply of oxygen was limited.

Consideration of the next method, that of precipita-

II—B. Lepsius, "Ueber die Abnahme der gelosten Sauerstoffs in Grundwasser and einen einfachen Apparat zur Entahme von Tiefproben in Bohrochern", Ber. Deut. Chem. Ges. 18, pp. 2487-2490, 1885.

III—Bull. U. S. G. S. No. 529, p. 90.

IV—Min. and Sci. Press, 106, p. 38, p. 141, p. 178, 1913.

V—H. V. Winchell, "Synthesis of chalcocite and its genesis at Butte, Mont. Bull. Geol. Soc. Am. 14, pp. 272-275.

VI—Bull. U. S. G. S., No. 529, p. 52.

tion by means of hydrogen sulfide formerly raised the question as to whether hydrogen sulfide can form in the zone of precipitation. The writer having established this fact that hydrogen sulfide can form, references to data on this point are omitted.

C. F. Tolman, Jr., and A. F. Rogers hold the opinion that many of our secondary copper deposits are in contact with primary ones and that the latter were brought up by alkaline solutions rich in carbon dioxide and hydrogen sulfide and that the escape or chemical assimilation of hydrogen sulfide has disturbed equilibria sufficiently to allow the precipitation of chalcocite from alkaline solution. The writer has shown that such action is possible and has shown the importance of hydrogen sulfide as a dispersing agent.

Cooke[1] was successful in precipitating silver sulfide by means of amorphous sulfur. Vogt[II] and Stokes[III] have shown that amorphous sulfur can readily form in nature.

One of the writer's most important conclusions deals with this action of amorphous sulfur.

Data on precipitation thru neutralization of de-

I—H. C. Cooke, "The secondary enrichment of silver ores", Jour. Geol. **21**, No. 1, p. 1.

II—J. H. L. Vogt, "Problems in geology of ore deposits," in Posepny, Franz. "The Genesis of Ore Deposits," pp. 676-677, 1902.

III—H. N. Stokes, Econ. Geol. **2**, pp. 14-23, 1907.

scending acidic solutions are rather numerous. E. C. Sullivan[I] secured a loss in copper content of such solutions by absorption or neutralization by keeping them in contact with powdered orthoclase albite, amphibole, shale, and clay gouge. Grout[II] has secured precipitation by neutralization with rather strong alkalies. His work is largely qualitative. W. H. Emmons[III] makes reference to the work of Ransome, Kemp and Lindgren, who show how limestone can be attacked by descending acid waters and can precipitate malachite and azurite in accordance with these equations:

$$2\ CuSO_4 + 2\ CaCO_3 + 5\ H_2O = CuCO_3.Cu(OH)_2 + 2\ CaSO_4 + 2\ H_2O + CO_2$$

$$2\ CuSO_4 + 3\ CaCO_3 + 7\ H_2O = 2\ CuCO_3.Cu(OH)_2 + 3\ CaSO_4 + 2\ H_2O + CO_2$$

The writer has secured some data on this subject of neutralization and its effect on precipitation.

Finally Lindgren, Graton and Gordon[IV] show chalcocite in coal, and give reduction by carbon as the cause of its deposition. Their equation is

$$4\ CuSO_4 + 5\ C + 2\ H_2O = Cu_2S + H_2SO_4 + 5\ CO_2$$

I—E. C. Sullivan, "The interaction between minerals and water solutions, with special reference to geologic phenomena". Bull. U. S. G. S. No. 312, pp. 37-64, 1907.

II—Econ. Geol. 8, p. 407-433, Aug. 1913.

III—Bull. U. S. G. S. No. 529, p. 101.

Reference and detail of work done are given under the writer's experiment No. 25.

CHEMICAL OUTLINE OF THIS INVESTIGATION AND REFERENCE TO GEOLOGICAL APPLICATIONS.

The results of this research, from the point of view of the chemist, may be considered as having to deal with two classes of metal-bearing solutions, (A) Solutions in which the copper is in the form of an electrolyte, and (B) Solutions of colloidal copper sulfides.

Taking up the matter covered by the first heading we come at once to oxidation. This was investigated early in the work. Failure to secure the solution of copper as a sulfate in more than traces, thru the agency of dissolved oxygen, (see exp. No. 1, No. 2, No. 3, No. 4 and No. 6) and great success in securing such when alternate drying and moistening were resorted to, (experiment No. 5), fully confirmed the views of Spencer[1] that dissolved oxygen probably cannot produce enough copper sulfate to form deposits of any consequence, and that in the zone of leaching we must have sufficient porosity to admit atmospheric air. The conditions under which the writer secured strong oxidation are those of the mining districts of the Southwest, where torrential rainfall is followed by long period of drought.

I—Econ. Geol. **8**, p. 631.

This work has shown that carbonate ores may, thru the agency of carbonated meteoric waters, go into solution or suspension and thus may readily contribute to the supply of copper in descending waters. (Experiments No. 7 and No. 8.)

Probably one of the earliest recorded syntheses of chalcocite is that of Winchell[11] and Tolman. They used acid copper sulfate solution with sulfur dioxide and secured a deposit on pyrite. The author has confirmed this result and has shown the great importance of impurities in the pyrite, (Experiment No. 9), and he has also established the fact that sulfur dioxide plays the role of a strong reducing agent, as thru its use he produced a crystalline cupro-cupric sulfite, (Experiment No. 10). In all solutions in which he used sulfur dioxide the precipitation of copper sulfide was greatly increased. (Experiment No. 20.) It is interesting to note in this connection that in the presence of ferrous sulfate, which generally has been thought to be an almost indispensable agent for the securing of chalcocite precipitation, the chalcocite is not as readily precipitated as in the absence of ferrous sulfate, but that the introduction of sulfur dioxide into such solutions containing ferrous sulfate, greatly increases this deposition. (Experiment No. 20). Acidity is shown to be less favorable than neutrality or alkalinity. As will be

---

11—H. V. Winchell, "Synthesis of chalcocite, etc." Bull. Geol. Soc. Am. 14, 269-276.

shown hydrogen sulfide is a powerful dispersing agent and is inimical to the deposition of the sulfides. Sulfide dioxide by its action on hydrogen sulfide, $2 H_2S + SO_2 = 2 H_2O + 3 S$, tends to overcome this dispersing action and in this manner also favors precipitation.

Doelter[1] is impressed with the fact that practically all substances are colloidal when first precipitated, yet that the crystalline condition seems to be the stable one in nature. He reviews the work and ideas of others and seems to conclude that what we call amorphous substances are very close to the crystalline. He speaks of the fact that the tendency to pass from the colloidal to the crystalline condition is greatly increased by pressure, shock, light and influence of Rontgen rays and radium rays and describes his own very successful work in producing crystalline material by means of shaking, long continued heating and particularly by means of pressure, and concludes that many minerals whose formation was once thought to require high temperatures may easily be formed without such temperatures. The ideas brought out in the paper seem particularly germane to the deposition and formation of massive chalcocite.

I—C. Doelter, "Ueber die Umwandlung amorpher Korper in Kristallinische", Zeit. f. Chemie und Industrie der Kolloide. Band 8, heft 1 s 29, heft 2, s 86.

Thru shock the cupro-cupric sulfite crystals mentioned above were produced by the author.

This fact that all precipitations are first colloidal was found to apply particularly to the sulfides of copper, and we now consider such colloidal solutions, our second main (B) topic of investigation. Under this is taken up: 1st, The formation of colloidal amorphous copper sulfides; 2d, The chemical conduct of amorphous copper sulfides, and, 3d, The physical conduct of amorphous copper sulfides.

This work shows clearly that colloidal copper sulfides can be formed from sulfate solutions and from the dispersion of crystalline or massive sulfides.

Taking up the formation of the sulfides from sulfate solution is was found that this could be brought about by any one of four methods.

Hydrogen sulfide, as is well known, is very efficient, and that it is quite possible geologically to have this hydrogen sulfide in the earth was shown in experiments No. 22 and No. 23. Hydrogen sulfide was produced by the action of water on pyrite and it was also produced by the action of very dilute acid on pyrite, chalcopyrite and bornite the relative development of hydrogen sulfide decreasing as the iron content of the minerals decreased.

Amorphous sulfur was found to precipitate a sulfide of copper from copper sulfate solution, this action being very rapid with fresh amorphous sulfur, prepared by the action of hydrogen sulfide on sulfur dioxide, freed from the gases and suspended in water, yet falling off rapidly as the sulfur aged. Experiment No. 19 shows this action, the curve being very striking. That sulfur should have this action is not at all surprising. The equation, $3 S + 2 H_2O = 2 H_2S + SO_2$ is known to be reversible. Stokes'[1] work readily accounts for the presence of amorphous sulfur in the zone of precipitation.

Several instances of production of amorphous copper sulfides by the action of sulfide minerals, particularly pyrite, on copper sulfate solution are shown in experiments No. 20, No. 21, and No. 35. Formerly this was thought to be *the one* method of sulfide production.

Finally, amorphous copper sulfides were produced by the action of a thiosulfate solution on one of copper sulfate. This action at first formed the double thiosulfate of sodium and copper which then decomposed to give a mixture of amorphous cupric and cuprous sulfides. Details of this work and references showing this action to be possible geologically are given experiments No. 18 and No. 24.

By the dispersion of crystalline and massive sulfides, the colloidal forms are very readily produced, and most effective for this dispersion is the agent hydrogen sul-

---

I—Econ. Geol. 2, 14-23, 1907.

fide. This disperses the crystalline or massive material in acid solution, (Experiment No. 43, No. 26, No. 27 and No. 28), and enormously more effectively in alkaline solutions, (Experiment No. 44, No. 38, No. 39, No. 40 and 41). The dispersion increases in a series chalcopyrite, bornite, covellite and chalcocite as the copper content of the minerals increases, which indicates a strong tendency for the sulfide of copper to migrate away from a sulfide of iron. Thus we see why chalcocite and chalcopyrite are so abundant, the former the one to migrate most and the latter to resist this action the longest.

It seems probable that the hydrogen sulfide forms an unstable compound with the copper sulfide. Linder and Picton[1] have shown that the dispersed material has a slightly larger sulfur content than the undispersed.

Free amorphous sulfur has the effect of dispersing a massive or crystalline sulfide. This is very forcefully shown in Experiment No. 13. In about four months a lump of chalcocite weighing seven grams lost over one gram in weight when sealed in a flask in contact with this amorphous sulfur.

Now, coming to the second part of our main heading (B) we may consider the chemical conduct of amorphous copper sulfides. Probably the most im-

---

[1]—Linder and Picton, "Some physical properties of arsenious sulfides and other solutions", Jour. Chem. Soc., **67**, p 63, 1895.

portant point in this investigation is concerned with the discovery of the fact that at ordinary temperatures cupric sulfide in contact with water spontaneously decomposes to cuprous sulfide and sulfur, $2\ CuS = Cu_2S + S$. (Experiments No. 31, No. 12, No. 15, No. 17, No. 18, No. 24, No. 26, No. 27, No. 29, No. 30, and No. 33.)

Fraulein Wassjuchnowa[II] showed that this action takes place in the dry state at temperatures above 500 degrees.

From the number of experiments cited one can see that this transformation took place very readily. It did not, however, take place as readily in the presence of ferrous sulfate as in its absence, (Ex. No. 14 and Ex. No. 16), and in those cases in which a large excess of free sulfur was used no large amount of Cu_2S was obtained in the length of time which the material stood. Cu_2S was also produced by use of sodium arsenite. This is described and discussed under experiment No. 32.

Finally coming to the physical conduct of the amorphous copper sulfides it is to be noticed that there is a tendency for them to accrete on certain minerals, and that they deposit from their dispersed condition when those agents which favor dispersion have their effects lessened. In experiment No. 13 it was found that

II—Zeitschr. f. Electrochemie **19**, No. 22, p 902, 1913.

some sulfide had gathered on chalcopyrite and in No. 29 a heavy dense growth was noted on bornite. Because of the relatively very much greater accretion on the bornite it would seem that this mineral may become chalcocite thru continued addition of copper sulfide, (See also Ex. No. 42). The same may be said of chalcopyrite tho in the case of this mineral the action would be very much slower. The fact that it is practically impossible to find a chalcocite which is free from specks of bornite gives additional weight to this view. This again is checked by microscopic evidence produced by Tolman and Ray[1].

The author was able to cause copper sulfide to accrete on sphalerite, (Ex. No. 30), and also on chalcocite as Cu_2S the phenomenon in this case being caused by the removal of a dispersing agent, H_2S.

The effect of the removal of this dispering agent H_2S is well brought out in Experiment No. 33. Here both crystalline, (apparently), and amorphous sulfides were thrown down as the hydrogen sulfide was released. Moreover with a reduction of alkalinity, such as in nature could be accounted for by descending acidic waters mingling with the alkaline solutions, a precipitate of the amorphous sulfide was produced.

With the complete elimination of hydrogen sulfide as is seen in experiment No. 41 and as seen in tendency

---

I—Unpublished manuscript.

in experiments No. 38 and No. 40, crystalline chalcocite was produced.

The above discussion throws some light on the question as to why chalcocite and chalcopyrite are the chief copper minerals of our mines.

This investigation was started with the view of determining why such should be the case, and it ends apparently with finding no reason why such should not be the case. All the processes involved in enrichment tend toward the formation of chalcocite. Only under particular conditions, (absence of hydrogen sulfide or presence of sulfur dioxide), would this be crystalline.

It is nevertheless evident that chalcocite, either crystalline or amorphous is the ultimate stable form toward which all copper-bearing minerals tend when in contact with solution. All other copper sulfide minerals are to be looked on as unstable. Of these chalcopyrite shows in all reactions a very low reaction rate which accounts for its apparently greater stability.

PREFACE TO EXPERIMENTS.

From the beginning of the work the author acted on no preconceived opinions but attempted to get data on any phase of the subject by means of any experiment which seemed to have a fair chance of being productive, and in all over sixty experiments were tried.

Each experiment was carried on carefully but with

attention to exactness commensurate with the expected results. This may be illustrated by the statement that the cyanide method was used in determining copper in some experiments, while in others gravimetric determinations were made on a very sensitive button balance. and all regeants were made up of selected crystals from C. P. analysed chemicals.

Duplication of the work will undoubtedly give closer figures at some points. It probably will alter no principle established here from experimental basis.

All experiments were carried on in the light and at ordinary laboratory temperature except as otherwise stated.

Experiment No. 1.—Study of oxidation of pyrite and of a pyrite and chalcocite mixture.

In belief that more light could be thrown on the subject of oxidation of pyrite and of pyrite and chalcocite mixtures, if such were in the form of very finely divided material which would be kept in water, this water being always saturated with oxygen, 100 grams of very pure pyrite which had been put thru a 200-mesh sieve were put into a long, narrow separatory funnel and covered with water. A tube was led into this and every third day oxygen was allowed to bubble thru the water for 24 hours.

At the end of the first 12 hours evidence of oxidation was seen. This continued to increase. Ferrous sulfate gathered in solution at the bottom of the cylinder and an "iron hat" formed at the top. This oxidation did not go on as fast as it had been expected to, and in view of other data obtained, (see Experiment No. 5), the passage of oxygen was discontinued, and the liquid was allowed to pass thru the pyrite and drain into a receptacle once each day. The drainings were poured back into the cylinder daily. *This treatment increased the oxidation.* At the end of 5 months the liquid contained 0.0393 g SO_4 and 0.0147 g H_2SO_4. The balance of the sulfate radical was distributed between ferrous and ferric iron.

A mixture of 50 grams 200 mesh pyrite and 200 mesh chalcocite[I] which had been treated in the same way, except for the passage of oxygen gave at the end of $5\frac{1}{2}$ months a solution which contained merely a trace of copper.

In view of the large amount of copper obtained by Buehler and Gottschalk[II], the idea suggested itself that as the cylinder was very long a precipitation of the copper sulfide might have taken place in the lower part of this cylinder. This was not investigated.

Experiment No. 2.—Study of oxidation in the presence of pyrite, manganese dioxide and solution of sodium chloride.

With the view of securing some data on the idea suggested by Lane[III], that copper minerals may go into solution in the form of chlorides, 50 grams of 200 mesh pyrite, 50 grams 200 mesh chalcocite and 0.5 gram 200 mesh manganese dioxide were put into a long, narrow separatory cylinder, covered with N/20

I—A very fine, exceedingly pure specimen furnished by Mr. A. C. Luhrs of Butte, Mont. Polished sections showed only mere traces of bornite.

II—Buehler and Gottschalk, "The Oxidation of sulfides" Econ. Geol. **5**, pp 28-35; **7**, pp 15-35.

III—A. C. Lane, Mich. Geol. and Biol. Survey. Pub. 6, Geol. Series 4, Vol. 2.

NaCl and treated with oxygen, etc., exactly as in Experiment No. 1.

At the end of 5½ months only a trace of copper had gone into solution, tho much manganese had gone into solution and had redeposited, evidently as hydrated oxide.

Experiment No. 3.—Study of solution of chalcocite.

A mixture of chalcocite and pyrite was treated as described in experiment No. 1. The water was kept saturated with oxygen for two months and then for 3½ months the tube was drained daily. Merely a trace of copper was found in solution.

Experiment No. 4.—[1]Attempt to secure copper in solution as a bicarbonate.

In the belief that oxidation would be very rapid if the minerals were very finely divided and if they were kept in water saturated with oxygen, 50 grams 200 mesh chalcocite, 50 grams 200 mesh pyrite and 50 grams of 60 mesh marble were put into a separatory funnel and treated as described in Experiment No. 1. Evidently the marble immediately used up any trace of sulfuric acid that was formed as no copper went into solution.

Experiment No. 5.—Study of oxidation of chalcocite.

Spencer[11] suggests that oxidation is most rapid where the ore is merely kept moist, and shows theoretically, at least, how the oxidation requires more oxygen than can be obtained from water solution.

Two glass tubes 12″ x 2″ were *nearly* sealed at each end. Thru the holes left open 10 grams of 200 mesh chalcocite was introduced into one, and 10 grams of

---

I—Tolman, "Secondary sulfide enrichment of ores" Min. and Sci. Press. **106**, p 180, 1913.

II—Spencer, Econ. Geol. **8**, p 631.

200 mesh chalcocite with 10 grams 200 mesh pyrite was put into the other. The mineral was covered with water and the tubes were set aside. These tubes were revolved once each day, thus the mineral was alternately moistened and dried.

At the end of only 29 days the liquid from the first tube showed a strong trace of copper. The liquid from the second tube showed 0.01896 g copper in solution.

Experiment No. 6.—Attempt to secure oxidation data.

In the belief that oxygen which could be liberated from hydrogen peroxide, would be liberated rather evenly thruout a mixture of powdered chalcocite and pyrite thus rapidly producing copper sulfate, a separatory funnel was prepared as described under experiment No. 1. The minerals acted as catalysers for the decomposition of the hydrogen peroxide, and at the end of a day or two about all the available oxygen had been set free. No trace of copper sulfate was secured.

Experiment No. 7.—Study of solution of malachite, azurite and chrysacolla.

As a result of many observations made in and around mines Professor C. F. Tolman, Jr., is of the opinion that these minerals can migrate downward, probably as bicarbonates.

Twelve gas washing bottles were taken. To each of 3 was added 30 cc N/2 potassium carbonate, to each of 3 others 30 cc potassium carbonate with a little sodium silicate, to each of 3 others 30 cc dilute sodium silicate solution, and to each of 3 others 30 cc of water, thus making 3 sets of 4 liquids in each set. I gram 200 mesh malachite was added to each bottle in one set, 1 gram 200 mesh azurite to each bottle in the second set and 1 gram 200 mesh chrysacolla to each in the third.

Carbon dioxide was passed thru each bottle for one

month. At the end of this time it was found that there was a trace of copper in the solution in each bottle, the liquids coming from the bottles containing chrysacolla and water, azurite and N/2 potassium carbonate and azurite and water giving the strongest tests.

Experiment No. 8.—Amount of copper going into solution or suspension from azurite and from chrysacolla.

To a gas washing bottle was added 1 gram 200 mesh chrysacolla and 100 cc water, to a second 1 gram 200 mesh azurite and 100 cc water and to a third 1 gram 200 mesh azurite and 100 cc N/2 potassium carbonate.

Carbon dioxide was passed thru each bottle very slowly for 96 days. The liquid in the first bottle was filtered and analysed. It contained 0.0004 g copper. The liquid in the second contained 0.0007 g copper and the liquid in the third 0.0007 g.

Experiment No. 9.—Synthesis of chalcocite. (Winchell-Tolman[1] experiment).

The experiment as described by these authors was repeated.

A lump of exceedingly pure pyrite was taken and placed into a solution of copper sulfate (1 cc = 0.019665 g Cu) whose acidity with sulfuric acid was N/20. This was then saturated with SO_2 and the whole was sealed in a bottle.

At the end of 18 days a faint trace of dark material appeared on some of the surfaces of the pyrite. At the end of one month one end of the pyrite was strongly coated while the other was as bright as ever. At the end of 5 months a uniform black coating covered the pyrite.

I—H. V. Winchell, "Synthesis of chalcocite etc" Bull. Geol. Soc. Am. 14, 269-276.

*Experiment No. 10.—Attempted synthesis of chalco-
cite.* (Winchell-Tolman experiment modified).

Because the tarnishing of the pyrite was so long de-
layed in Experiment No. 9 it was believed that impurities were necessary for the formation of chalcocite, and
as the analysis of the pyrite used by Winchell and Tolman showed the presence of zinc, a crystal of sphalerite
was wired to a piece of very pure pyrite with platinum
wire. This couple was then put into a solution such as
was used in No. 9 and the flask was sealed.

At the end of 46 days no marked change could be
seen. On the 47th day the flask was accidentally subjected to great jarring and on the 48th day small ruby
red crystals were seen. These were prismatic, had
high relief, were pleochroic and had parallel extinction.

As they gave all the tests for cupro-cupric sulfite as
described by Segerbloom in his "Table of Properties"
they were undoubtedly cupro-cupric sulfite.

This experiment confirms the belief of Winchell[I]
and Spencer[II] that SO_2 plays the role of a strong reducing agent. It also tends to confirm the opinion of
Lindgren[III] that pyrite does not precipitate Cu_2S or
CuS while zinc blende is present. Neither sulfide
formed after months of standing.

*Experiment No. 11.—Effect of jarring on crystalla-
tion.*

The experiment as described under No. 10 was repeated. The sealed bottle was allowed to stand for
nearly double the length of time which the bottle stood
in No. 10. No crystals appeared. The bottle was then

I—Bull. Geol. Soc. Am. 14, pp 273-275.

II—A. C. Spencer, "Chalcocite deposition" Jour. Wash. Acad.
    Sci. 3, pp 70-75, 1913.

III—Lindgren etc. Prof. Paper. U. S. G. S. No. 43, p 183, 1905.

subjected to considerable jarring and crystals appeared within a few hours.

Experiment No. 12.—Production of cuprous sulfide.

H_2S was passed thru a solution of $CuSO_4$ until all of the copper had precipitated as CuS. The liquid was filtered off and the precipitate was surrounded with H_2S water, and sealed in a flask with an atmosphere of H_2S.

At the end of 24 hours a ring of what was, from its color, covellite streaked the flask at the surface of the precipitate. At the end of a few days this covellite was not noticeable.

The flask was opened at the end of 130 days. The residue showed an abundant separation of sulfur when examined with the microscope.

This residue was washed, dried, extracted with CS_2 and 0.7173 g was taken for analysis. This gave 0.1723 g sulfur thus leaving 0.5450 g copper.

0.7173 g Cu_2S contains 0.5729 g Cu.

0.7173 g CuS contains 0.4767 g Cu.

On the basis of the copper content this residue contained 70.99% Cu_2S.

Experiment No. 13.—To observe transference, substitution or growth.

A lump of chalcocite weighing 7.6332 g and a lump of chalcopyrite weighing 3.2458 g were sealed in a flask with 50 cc copper sulfate solution, (1 cc $=$ 0.039331 g Cu), 50 cc ferrous sulfate solution, (1 cc $=$ 0.026243 g Fe), and a large excess of amorphous sulfur, (2 days old) and an atmosphere of carbon dioxide.

A decided growth of black material was observed on and in contact with the chalcocite at the end of two weeks.

The flask was opened at the end of 137 days. The

chalcocite lump remained attached to the bottom of the flask. This lump was covered with a black velvety film which stood out in ridges on the lump. The film, (very loosely attached), was removed and the lump was weighed. It weighed 6.5745 g—a loss of 1.0587 g.

This film and the general residue in the flask were analysed. Both contained iron which made exact results as to Cu_2S and to CuS of little value. The fact that the film had a copper content of 69.03% Cu, while the general residue contained 64.73% showed the greater copper content to be nearer the lump of chalcocite.

It seems probable that in the presence of the excess of amorphous sulfur that the sulfide was mostly CuS tho some Cu_2S was probably present in the film.

The chalcopyrite lump was loose and came out of the flask readily. It weighed 3.2473 g—a gain of 0.0015 g. A deposit of black sulfide hung tenaceously to the lump and could not be washed off. This would account for the gain, and would indicate that chalcocite can grow on or at the expense of chalcopyrite, but not as readily as on or at the expense of bornite. (See Experiment No. 29).

This experiment was repeated with amorphous sulfur which was one day old, and similar results were obtained. It was also repeated with freshly prepared amorphous sulfur. The final results were much as given above.

Experiment No. 14.—To observe transference, substitution or growth.

As described under No. 13, a lump of chalcocite and a lump of chalcopyrite were put into a solution of copper sulfate mixed with one of ferrous sulfate and sealed in an atmosphere of CO_2.

At the end of 76 days the chalcocite showed a loss of 0.0817 g. The weight of the chalcopyrite was unchanged.

This agrees with the results obtained by Emmons and Grout[IV]. They secured no chalcocite deposition by using ferrous sulfate as a reducing agent.

Experiment No. 15.—Production of cuprous sulfide.

Cupric sulfide was washed free from electrolyte and was added to water which had been saturated with H_2S. The whole was sealed in a flask with an atmosphere of H_2S.

It was very noticeable that in the absence of an electrolyte the CuS remained colloidal and dispersed, and for 52 days so remained—no clear liquid appearing above the precipitate. On the 53rd day it was noticed that a clear layer had appeared. In a few days the precipitated was well compacted and lay on the bottom of the flask.

At the end of 75 days the flask was opened. The black precipitate was seen with the miscroscope to be a mass of well-compacted particles of copper sulfide which were mixed with particles of sulfur.

This precipitate was washed with water and alcohol and then *partly* extracted with CS_2. The CS_2 gave much sulfur upon evaporation.

0.3921 g of the residue was analysed for Cu and for S. It gave 0.1266 g S and 0.2655 g Cu. 0.3921 g Cu_2S yields 0.3178 g Cu and the same weight of CuS yields 0.2606 g Cu. On the basis of the copper content of the residue there was present 8.57% Cu_2S. It probably contained a larger percentage.

Experiment No. 16.—Attempt to produce cuprous sulfide.

IV—W. H. Emmons, Bull. 529. U. S. G. S. p 108.

100 cc of copper sulfate solution and 100 cc of ferrous sulfate solution (1 cc equalling 0.33931 g Cu and 1 cc equalling 0.02624 g Fe respectively) were mixed, saturated with H_2S and were sealed in a flask with an atmosphere of H_2S.

At the end of 9 days miscroscopic particles of sulfur could be seen on the side of the flask. At the end of 120 days the contents of the flask were filtered and the residue was analysed. The residue was found to contain both cuprous and cupric sulfides.

When the filtrate was neutralized with NaOH a black precipitate of copper sulfide came down with the $Fe(OH)_2$.

Experiment No. 17.—Production of cuprous sulfide.

100 cc of a solution of copper sulfate, (1 cc $=$ 0.039331 Cu), was saturated with SO_2, then with H_2S, then with SO_2 and the whole sealed in a flask with an atmosphere of SO_2.

At the end of 11 days the flask broke *inwardly*. The material as transferred to a new flask and given a new atmosphere of SO_2 and was sealed.

At the end of 58 days a lump raised above the surface of the precipitate to the height of one-half inch and caused the precipitate to appear as if a lump of ore were embedded in it.

This lump continued to grow, and another came. From the first lump some elevated points, suggestive of horns, struck well into the solution. When the flask was finally opened the lumps being very fragile were destroyed. One of the "horns" remained whole and under the microscope was seen to be a large prism. The residue in the flask consisted of black copper sulfide and sulfur.

When, at the end of 136 days, the flask was opened the odor of SO_2 could not be observed.

The residue was analysed after having been washed, extracted with CS_2, etc. 0.7065 g gave 0.1714 g S thus leaving 0.5351 g Cu. 0.7065 g of Cu_2S contains 0.5643 g Cu while the same amount of CuS contains 0.4695 g Cu. On the basis of the copper content the residue was 69.16% Cu_2S.

The clear supernatant liquid in the flask was filtered and a lump of zinc was added to the filtrate. A very delicate brownish black cloud gathered in the vicinity of the zinc. This was considered to be a sulfide of Cu, (it gave a strong Cu test), which had been held in solution or suspension by H_2S, (derived probably from S), and that when this H_2S was removed by the partial pressure of the H from the zinc, or the sulfide was precipitated by the electrolytic action of the zinc, this cloud appeared.

Experiment No. 18.—Production of cuprous sulfide by the use of sodium thiosulfate.

The rapid precipitation of Cu by amorphous S led to the suggestion of attempting to get this S from the decomposition of a thiosulfate as Hillebrande[1] seems to have found thiosulfates in deep mine waters, and as Stokes[2] had shown that this is produced when alkaline solutions act on pyrite.

Three flasks were taken and to each was added 50 cc $CuSO_4$ solution, (1 cc = 0.039331 g Cu). To the first was added 5 cc N/2 $Na_2S_2O_3$, to the second 25 cc and to the third 50 cc of this thiosulfate solution. Each

---

I—17th Ann. Rep. U. S. G. S., part 2, p 21, 1896.

II—H. N. Stokes, "Experiments on the action of various solutions on pyrite and marcasite". Econ. Geol. 2, pp 14-23, 1907.

flask was given an atmosphere of CO_2 and was sealed.

In each flask the double thiosulfate of sodium and copper formed at once and then began to decompose. The contents of the first flask were black in 2 days, in the second flask in 5 days and in the third at the end of several weeks.

The formation and decomposition of the double thiosulfate of sodium and copper, which according to Dutoit[III] has the copper in the cuprous condition, goes in several steps or stages. When the two solutions are mixed tiny yellow prisms are formed. These on standing produced long yellow needles which begin to splinter, even going so far as to be a mass of radiating splinters, and these splinters decompose to a black sulfide, decomposition first taking place at the end of the splinters.

At the end of 71 days the residue in the second flask was washed with water and with alcohol, and was extracted with CS_2 and analysed. 0.5035 g of the residue gave 0.1563 g S thus leaving 0.3472 g Cu.

0.5035 g Cu_2S contains 0.4022 g Cu and this amount of CuS contains 0.3347 g. On the basis of the copper content this residue was 21.74% Cu_2S.

Experiment No. 19.—The effect of age on the precipitating power of amorphous sulfur.

Early in this work it was noticed that freshly prepared amorphous sulfur precipitated copper sulfide from solutions of copper sulfate much more readily than the older material.

In order to bring out this fact somewhat quantitatively H_2S and SO_2 were brought together to produce amorphous sulfur. This when free from the gases was

III—Pierre Dutoit, "Sur les hyposulfites cupro alkalins" Jour. de Chemie Physique **2**, 4, pp 650-673, 1913.

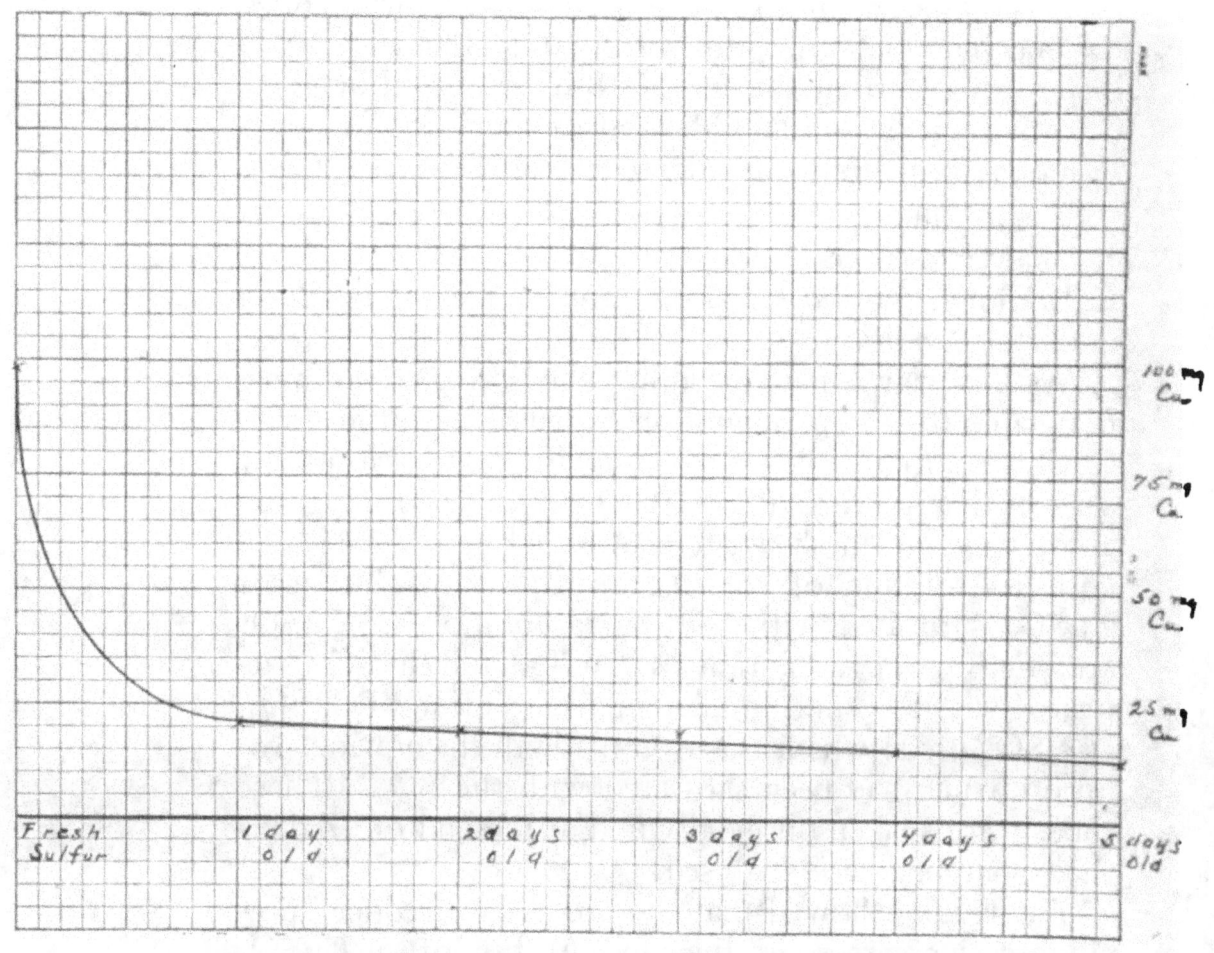

Curve showing that as amorphous sulfur ages it can precipitate smaller amounts of copper sulfide from a sulfate solution in a given amount of time.

put into suspension with water and more than enough of this sulfur to precipitate a given amount of copper from a sulfate solution was added to each of seven bottles, the same amount being added to each bottle. A given amount of copper sulfate solution was added to one bottle at once, to a second at the end of 24 hours, to a third at the end of 48 hours, etc. Each bottle was then shaken and allowed to stand for 24 hours when its contents were filtered and analysed.

The results of the experiment are shown in the accompanying curve.

Experiment No. 20.—Precipitating power of pyrite in various solutions.

With a view of ascertaining those conditions most favorable for the precipitation of Cu as a sulfide from $CuSO_4$ solutions by means of pyrite, and with a view of securing more data on the role of SO_2 a given amount of 200 mesh pyrite was sealed in tubes with an aqueous solution of $CuSO_4$, an aqueous solution of $CuSO_4$ saturated with SO_2, in a solution of $CuSO_4$ and $FeSO_4$, in a solution of $CuSO_4$ and $FeSO_4$ saturated with SO_2, and in an acidified solution of $CuSO_4$, (H_2SO_4 used) which was isotonic with the aqueous solution which had been saturated with the SO_2. These tubes were shaken once daily and at the end of 47 days their contents were analysed.

The results showed that the amounts of copper precipitated increased with the concentration of the $CuSO_4$ solution, that great precipitation takes place in aqueous solution, and that this is increased by the presence of SO_2. The results also showed that the precipitation in the presence of $FeSO_4$ is not as great as in its absence, but that SO_2 increases this precipitation in $FeSO_4$. They showed that the reducing action of SO_2 and not the action of the H ion is responsible for the increased precipitation in the presence of this substance. Moreover they showed that neutrality or alkalinity are conditions most favorable for precipitation.

Details are given in the following table:

	1 Amt. Cu ppted from water solution of CuSO₄	2 Amt. Cu ppted from water sol. of CuSO₄ sat. with SO₂	3 Amt. Cu ppted from approx. N/10 FeSO₄ sol. of CuSO₄	4 Amt. Cu ppted from approx. N/10 FeSO₄ sol. of CuSO₄ sat. with SO₂	5 Amt. Cu ppted from H₂SO₄ sol. of CuSO₄ isotonic with sol. in column 2
1 g pyrite 22 cc liquid 0.0787 g Cu as CuSO₄	0.01325	0.0240	0.0113		negligible
1 g pyrite 24 cc liquid 0.1573 g Cu as CuSO₄	0.0269	0.0446	0.0187	0.0119*	"
1 g pyrite 26 cc liquid 0.2360 g Cu as CuSO₄	0.0277	0.0677	0.0250	0.0295	"
1 g pyrite 28 cc liquid 0.3146 g Cu as CuSO₄	0.0364	0.0972	0.0318	0.0462	"

Experiment No. 21.—Precipitating power of mono and disulfides and of mono and diarsenides.

With a view of ascertaining whether disulfides and diarsenides would precipitate more copper than the mono compounds, a series of minerals were taken, any

* Probably experimental error.

impurities removed mechanically as far as possible, and the minerals were put thru a 200 mesh sieve. One gram of each mineral was sealed in a tube with an aqueous solution of $CuSO_4$, each tube shaken daily and at the end of 47 days the contents were analysed.

As a result of this experiment it was concluded that no generality was shown. Details are given in the table.

	Cobaltite	Niccolite	Niccolite	Chloanthite	Enargite	Enargite	Bornite	Galena	Alabandite	Stibnite	Pyrrhotite
Amt. Cu in 25 cc solution	.0916	.0916	.1374	.1374	.0916	.1374	.0916	.1374	.0916	.0916	.0916
Amt. Cu in filtrate	.0774	.0658	.1089	.0415	.0825	.1262	.0820	.1040	.0000	.0692	.0726
Amt. Cu. ppted	.0142	.0258	.0285	.0959	.0091	.0112	.0096	.0334	.0916	.0224	.0190

Experiment No. 22.—Hydrogen sulfide produced by hydrolytic action of water on pyrite.

Several one-half gram portions of uniform sized pyrite particles, ranging from those which passed thru the 20 mesh sieve and were retained on the 30 mesh to those which passed thru the 200 mesh sieve were sealed, each portion in a separate tube, with 20 cc of water.

These tubes were kept at about 41 degrees for three months, when the tube containing the 80-90 mesh pyrite was opened, its contents filtered and tested colorimetrically with a standard solution of lead nitrate. The test showed the presence of H_2S in solution. Other tubes were afterward tested and their contents gave like results.

Experiment No. 23.—Production of hydrogen sulfide from pyrite, chalcopyrite and bornite.

Duplicate tubes containing:

(a) 0.5 g 200 mesh pyrite and 15 cc N/10 sulfuric acid

(b) 0.5 g 200 mesh chalcopyrite and 15 cc N/10 sulfuric acid

(c) 0.5 g 200 mesh bornite and 15 cc N/10 sulfuric acid

were sealed and placed in a thermostat at about 41 degrees for 51 days.

The contents of one set of tubes were filtered and the filtrates treated with standard lead nitrate solution. All showed the presence of H_2S in solution, the pyrite having set free the most and the bornite the least.

The duplicate set was then put in an autoclave and heated at 195 degrees for two hours. The tube containing the acid and pyrite exploded, the one containing the acid and chalcopyrite cracked and the one containing the acid and bornite remained intact, tho much pressure was developed by the H_2S inside. This H_2S was readily noticed because of its odor.

Experiment No. 24.—Cuprous sulfide from the double thiosulfate of sodium and copper.

Some of the double thiosulfate was prepared from solutions of copper sulfate and sodium thiosulfate. This was washed and sealed in a flask with an excess of water and an atmosphere of CO_2.

After 8 weeks the material in the flask began to gather into large platy masses which stood up on one edge in the solution. This continued until the material in the flask was well compacted into large platy masses.

The flask was opened at the end of 123 days, and the

residue was washed with water, alcohol and was dried and extracted with CS_2 and analysed.

0.5170 g of the residue gave 0.1380 g S thus leaving 0.3790 g Cu. 0.5170 g of Cu_2S contains 0.4130 g Cu, while this weight of CuS contains 0.3437 g Cu. On the basis of the copper content of this residue it contained 50.94% Cu_2S.

Experiment No. 25. — Attempt to precipitate copper sulfide by the use of coal.

Lindgren, Gratton and Gordon think that coal has acted as a reducing agent on copper solutions and in their "Ore Deposits of New Mexico" show a beautifully colored plate of chalcocite embedded in coal.

With a view of securing data on this method of precipitation 2 grams of powdered coal were put into a solution of $CuSO_4$ and $FeSO_4$ and the whole was saturated with SO_2 and sealed in a flask. Some action was expected in the presence of these three reducing agents.

The flask was shaken daily and at the end of 122 days was opened and its contents were examined. No deposition of sulfide of copper was observed.

Great care was taken to secure a coal which had no pyrite in it. The results of this experiment seem to indicate that carbon alone will not precipitate the sulfide. Had pyrite been present in the coal some sulfide would undoubtedly have been precipitated.

Experiment No. 26.—Production of cuprous sulfide.

A lump of chalcocite weighing 2.2302 g was put into a solution of copper sulfate, (1 cc = 0.039331 g Cu), and this in a flask was connected with an H_2S generator so that the H_2S was always present in the flask.

I—U. S. G. S. Prof. Paper No. 68, p 57.

The H_2S was not passed thru the liquid but was allowed to settle upon it.

At the end of 4 days what appeared to be shining crystalline faces were seen on the lump of chalcocite, and in spite of the precipitation of some CuS from solution the liquid appeared to be more blue than ever. This latter phenomenon was attributed to the reflected color of covellite films.

At the end of 100 days the flask was opened. The chalcocite was found to weigh 1.9923 g—a loss of 0.2379 g. The lump was badly pitted.

A film had gathered around the lump. 0.1083 g of this was taken for analysis. It was not extracted with CS_2. It gave 0.0299 g S thus leaving 0.0784 g Cu.

0.1083 g Cu_2S contains 0.08648 g Cu while the same weight of CuS contains 0.07197 g Cu. On the basis of the copper content of the film, this contained 43.1% Cu_2S.

Hard films of the CuS which had originally formed in the flask could be separated from the general residue. These were washed, extrated and analysed. 0.4718 g gave 0.1437 g S thus leaving 0.3281 g Cu. 0.4718 g Cu_2S contains 0.3768 g Cu and the same weight of CuS contains 0.3135 g Cu. On the basis of the copper content this film was 23.07% Cu_2S.

The general residue in the flask was treated as described above and analysed. 0.6522 g gave 0.2002 g S and 0.4520 g Cu. Thus the Cu_2S content of the general residue was 21.28%.

It is to be noted that the chalcocite lump lost weight, that the film on the lump was richer in Cu_2S than any other part of the material in the flask, and that the residue near the film was richer in Cu_2S than the general residue. Thus we see the tendency of the Cu_2S to give up its Cu to nearby CuS, (the equilibrium point we do not know), and the tendency of the CuS to drop off its S and become Cu_2S, or because of the dispersion of the chalcocite the material nearest to the lump became most contaminated with Cu_2S.

Experiment No. 27.—Production of cuprous sulfide.

A lump of pyrite weighing 2.7769 g was placed in a flask and kept under conditions as described under experiment No. 26.

At the end of one week a color and luster resembling covellite was noticeable.

At the end of 91 days the flask was opened. The pyrite weighed 2.7755 g—a loss of 0.0014 g.

The residue was washed, dried, extracted, etc. 0.9059 g gave 0.1994 g S thus leaving 0.7065 g Cu. 0.9059 g Cu_2S would give 0.7235 g Cu and the same weight of CuS would give 0.6020 g Cu. On the basis of the copper content this residue contained 86.02% Cu_2S.

Experiment No. 28.—Production of cuprous sulfide.

Under conditions as described for Experiment No. 26, 1.8291 g covellite was placed in contact with the sulfate solution and the H_2S. It cannot be stated just how long the H_2S acted on the solution as its supply was accidentally interrupted.

At the end of 100 days the flask was opened. The covellite weighed 1.8241 g—a loss of 0.0050 g.

The residue when treated as described showed a content of 53.19% Cu_2S.

Experiment No. 29.—Production of cuprous sulfide and growth on bornite.

Under conditions as described under experiment No. 26, 4.2770 g bornite was placed in a flask with the copper sulfate solution and connected with the H_2S generator.

At the end of one week the color and luster of covellite was noticeable.

At the end of 94 days the flask was opened. The lump of bornite was found to weigh 5.5569 g—a gain of 0.2799 g. This growth was noticed largely in the gain in weight as the material which had been incorporated with the lump was so added that the appearance of the lump was hardly altered, except that it was a deeper dark blue than usual. This experiment seems to indicate that the copper sulfide tends to concentrate around the bornite. Thus bornite could turn to chalcocite thru a gradual increase in its copper content.

On the same basis (e. i., copper content) as stated in Experiment No. 26 the residue in the flask was found to be 16.62% Cu_2S.

Experiment No. 30.—Production of cuprous sulfide.

A crystal of sphalerite weighing 6.7795 g was placed in a flask under conditions as described in experiment No. 26.

At the end of one week the color and luster of covellite was noticeable.

At the end of 93 days the flask was opened. The lump of sphalerite was found to weigh 6.7745 g—a loss of 0.0050 g, yet a black deposit of copper sulfide hung tenaceously to the sphalerite along crystallographic lines. The whole crystal was not covered with the deposit. It seemed that the sphalerite is an excellent precipitant of the sulfides of copper[1] yet that in the case of this particular sphalerite crystal only a portion

---

1—Prof. A. F. Rogers of Stanford University has produced covellite by the simple heating of a solution of $CuSO_4$ with sphalerite in a bomb furnace. School of Mines Quarterly, 32, p 298, 1911.

of the precipitated sulfide remained on the sphalerite.

The residue in the flask was treated and analysed as has been already described. It showed a Cu_2S content of 29.57%.

Experiment No. 31.—Production of cuprous sulfide.
    (This is the experiment in which the spontaneous change of CuS to Cu_2S and S was first noted.)

A solution of $CuSO_4$ had H_2S passed thru it until all the copper was precipitated as the sulfide. The precipitate was freed from solution by filtration and then this precipitate was placed in a flask with a lump of chalcocite weighing 2.2861 g and the whole was covered with water and sealed an an atmosphere of CO_2.

After 54 days the flask was opened. The lump weighed 2,1404 g—a loss of 0.1457 g. The residue under the microscope was seen to be made up of particles of black sulfide which were intermingled with particles of sulfur. The residue was extracted with CS_2 (which was known to be free from dissolved S) and the extract evaporated. A yellow residue which by the miscroscope and by chemical test was shown to be sulfur was left upon evaporation of the CS_2.

The residue from which the sulfur had been extracted was analysed for both S and Cu, and upon the basis of its Cu content was shown to be 43.64% Cu_2S. Thus the transformation of CuS to Cu_2S was shown.

Experiment No. 32.—Production of cuprous sulfide.

A suggestion that sodium arsenite can transform CuS to Cu_2S[I] led to the sealing of 15 g CuS and 30 g sodium arsenite in a flask with an atmosphere of CO_2. 200 cc H_2O was present.

After 55 days the flask was opened and its contents

I—Gmelin-Kraut "Handbuch der anorganische Chemie" Vol. 5, p 807.

were analysed. They showed a Cu_2S content of 49.00%.

This transformation was at first attributed to the sodium arsenite but in view of spontaneous changes obtained in other experiments the amount to be attributed to each cause is problematical.

Experiment No. 33.—Growth of cuprous sulfide on chalcocite.

With a view of getting some experimental cerification of the formation of chalcocite as suggested by Prof. A. F. Rogers[II] a lump of chalcocite weighing 4.5240 g was put into a flask containing CuS and a solution of K_2S which had been completely saturated with H_2S. Testing showed that this liquid held considerable Cu in suspension or solution. A stopper with a Bunsen valve was inserted, thus giving the H_2S which held the copper in suspension, a chance to escape.

At the end of three weeks crystal faces were seen shining on the chalcocite lump.

At the end of 85 days the flask was opened and the solution displaced by water. The chalcocite lump was washed with hot water until it was free from alkali. One could see that the lump was coated and that this coating had on it tiny shining spots suggestive of crystal faces. The lump was dried. It weighed 4.5874 g— a gain of 0.0634 g. 32.24 milligrams of this coating were taken for analysis. This gave 8.403 mg of S, thus leaving 23.837 mg of Cu. 32.24 mg of an absolutely pure chalcocite would give 25.748 mg Cu.

This sulfur figure was known to be too high. While C. P. chemicals were used a very noticeable amount of iron oxide was detected in the $BaSO_4$ precipitate. Furthermore, as no extraction by CS_2 was made it is quite

---

II—A. F. Rogers, Econ. Geol. **8**, p. 781, 1913.

possible that a small amount of S may have been in the material deposited on the chalcocite.

The author has no hesitation in saying that he produced a pure chalcocite from this alkaline solution.

The analysis of the residue in the flask showed that it had a Cu_2S content of 57.63%.

Experiment No. 34.—Attempt to secure miscroscopic evidence of the change of covellite to chalcocite.

H_2S was passed thru a solution of $CuSO_4$ until all the copper was precipitated as a sulfide, then some 200 mesh chalcocite was added and the whole was sealed in a flask with an atmosphere of CO_2.

A similar set up was made with 200 mesh covellite.

After 23 days both flasks were opened. Their contents showed in each case that the powdered mineral was embedded in the amorphous sulfide. Free sulfur was noted in each case. Both flasks were again sealed with CO_2 atmospheres and both were opened again at the end of 113 days. The appearances of their contents were much as before, except that a much larger amount of free sulfur could be seen.

Experiment No. 35.—Attempt to find the condition of acidity or alkalinity most favorable for the precipitation of copper sulfide from copper sulfate solutions by means of pyrite.

One gram of 200 mesh pyrite was put into each of 11 tubes, a given amount of $CuSO_4$ was added to each and then such amounts of water and H_2SO_4 were added as to have the acidity of the solutions in the different tubes as follows: N/15, N/20, N/40, N/80, N/100, N/200 and "neutral". To other tubes NaOH and water was added so that after the amount of NaOH had been added which would theoretically throw down all of the

Cu as $Cu(OH)_2$ the alkalinity of the tubes was N/200, N/100, N/80, N/40.

The contents of those tubes containing the alkali were black at the end of 24 hours.

Each tube was shaken once daily and after 31 days the contents were analysed.

It was found that under condition of acidity there was *practically no copper sulfide precipitated.*

In the case of the tubes containing the alkali the difficulty presented itself of finding out how much of the black material was copper oxide and how much was a sulfide of copper. It seemed reasonable to believe that as the sulfide precipitates in acidity of 1:8 HCl that if the residue were washed with a 1:10 HCl the sulfide would not be dissolved. This proceedure was taken. All of the material dissolved, thus this experiment proved that acidity was not favorable to precipitation by means of pyrite but left the question as to alkalinity unsettled. (See also Experiment No. 20.)

Experiment No. 36.—Neutralizing power of different rocks.

As altered rock is always found in veins, etc., showing *downward* enrichment, and as this alteration is particularly kaolinization, and as experiments No. 20 and 35 indicated that acidity is not favorable to the deposition of the sulfides a series of typical rocks was taken and the neutralizing effect of each rock on acid solution was determined. Each rock was put thru a 200 mesh sieve. 5 grams of each was put into a bottle and covered with 50 cc N/10 sulfuric acid. The bottle containing each was shaken once each day. After 30 days the liquid in each bottle was analysed.

Details are given in the accompanying table:

Rock and Locality.	Principal minerals in order of abundance	Amt. N/10 H_2SO_4 used	Amt. N/10 H_2SO_4 neutralized	Relative neutralizing power of the rock, granite being taken as 1.
Franciscan Limestone. Back of Stanford University	Calcite Chalcedony	50 cc	48.5 cc	10.32
Basalt. Stanford University Quarry.	Plagioclase Augite Orthoclase Zeolites Pyrite	50 cc	23.4 cc	4.98
Diorite. Trinity River. So. Fork, near Low Gap, Cal.	Orothclase (serecitized) Hornblende Biotite Epidote Chlorite	50 cc	16.7 cc	3.55
Franciscan Shale. Telsa, Cal.	Clay substance Quartz Fe bearing minerals	50 cc	15.1 cc	3.21
Rhyolite. Alum Rock, San Jose, Cal.	Glass (devitrefied and silicified) Quartz and Chalcedony Orthoclase Plagioclase Kaolin Iron stain	50 cc	11.0 cc	2.34
Hornblende Andesite. Marysville, Cal.	Glass Hornblende Plagioclase Orthoclase	50 cc	11.0 cc	2.34
Granite. Santa Lucia, Cal.	Orthoclase Plagioclase Quartz Biotite Apatite	50 cc	4.7	1.00

Experiment No. 37.—Attempt to produce cuprous sulfide.

A lump of pyrite weighing 4.2663 g and some carefully washed CuS were put into a solution of N/2 K_2CO_3 and the whole was sealed in a flask with an atmosphere of CO_2.

At the end of 95 days the flask was opened. The pyrite was found to weigh 4.2653 g—a loss of 0.0010 g. The pyrite was bright and shiny.

The residue when examined with the miscroscope showed that some sulfur had separated. This residue was lost in the process of analysis. Judged from its appearance, however, this residue was made up of CuS with a small amount of Cu_2S.

Some iron was observed in the liquid contained in the flask.

Experiment No. 38.—Attempt to produce cuprous sulfide in alkaline solution.

CuS and a fairly concentrated solution of K_2S were heated for some time, finally at the boiling temperature of the solution. The liquid was allowed to settle until clear. The clear supernatant solution was placed in a dessication tube over sulfuric acid, and this tube was evacuated and sealed.

At the end of 24 hours a brownish mass had formed. This with the hand lense had the appearance of a lot of tiny needle-like crystals. With high magnification these were seen to be long curved brownish hairs. When viewed with polarized light these hairs failed to show crystalline characteristics. Some sulfur could be seen along with these hairs.

The liquid in the dessication tube constantly grew more concentrated as the sulfuric acid took up the water vapor and at the end of 113 days these were seen to be masses of amorphous copper sulfide, probably cuprous as other experiments showed.

*Experiment No. 39.—Solution and deposition of chal-
cocite.*

Some 200 mesh chalcocite was put into a strong solution of KOH and H_2S was passed thru the solution,
which was kept cold, until the solution was saturated.
The heavy dense chalcocite became collodial, occupying
4 or 5 times its original volume. This whole solution
was heated in an autoclave at 180 degrees for 1 ½ hours.
The solution became deep yellow. This was filtered
thru glass wool and was then gently heated on an air
bath. H_2S came off slowly and finally black lumps
gathered on the bottom of the container.

A drop of this filtrate just mentioned was evaporated
on a glass slide and its residue under the microscope
showed black needles or prisms together with some yellow prisms suggestive of the double thiosulfate of sodium and copper. Some of this same filtrate was neutralized with dilute sulfuric acid whereupon a brownish black precipitate was formed. This gave a strong
test for copper.

*Experiment No. 40.—Solution and deposition of a cop-
per sulfide derived from covellite, bornite and chalcopyrite.*

Experiment No. 39 was repeated with each of the
minerals named above.

The filtrate from covellite behaved as did the filtrate
obtained in No. 39. A few of the hairs as obtained in
Experiment No. 38 were noted on the glass wool filter.

The filtrate from the bornite behaved as did that
from the covellite. It did not contain as much copper,
however. A larger number of the hairs were obtained
from the bornite. These gave a very strong test for
copper.

The filtrate from the chalcopyrite gave only a faint test for copper.

It appeared that the H_2S had dispersed these minerals in the following descending order: chalcocite, covellite, bornite and chalcopyrite and that the alkaline liquid with its H_2S had dissolved copper from these minerals in this same order.

The dispersion and solution of the chalcocite was much more marked than were these phenomena with the other minerals.

Experiment No. 41.—Production of chalcocite crystals, and the effect of hydrogen sulfide as a dispersing agent.

2 grams of 200 mesh chalcocite were put into a tall Nessler tube and covered with 25 cc of a KHS solution. H_2S was passed thru the solution for an hour.

The same amount of chalcocite was put into a tube and covered with 25 cc of a K_2S solution. Hydrogen was passed thru this solution for an hour.

The same amount of chalcocite was put into another tube and was covered with 25 cc of a solution of equal parts of K_2S and KOH. Hydrogen was passed thru this liquid for an hour.

Each solution contained approximately the same amount of potassium. The gases were passed thru the liquids at approximately equal rates.

The material in the first tube showed strong dispersion, that in the second showed 1/4th to 1/6th as much and that in the third showed scarcely any.

At the end of 10 days it was noted that the material in the second tube had risen, and that a small mass having something of a crystalline appearance floated on the surface. From this hung down a growth resembling a fox's tail.

This growth and "tail" were washed, and the latter examined with a microsscope. This "tail" was seen to be made up of a lot of stiff black, non-transparent hairs or rods. One or two were especially large and showed truncated ends. The angles on the ends of one rod were measured. Beginning on the right side of the rod or prism and going anti-clock wise the angles were approximately 37, 51 and 56 degrees. (See figure.)

The floating mass was washed again with water and with absolute alcohol and was then dried. It was found to weight 13 mg. It was analysed with the greatest care and was found to yield 2.57 mg S and 10:40 mg Cu. 13 mg of Cu_2S would yield 10.39 mg Cu. The floating mass was certainly cuprous sulfide. The author assumes that the crystals were of the same composition.

Experiment No. 42.—Elimination of iron from bornite.

Work done simultaneously with this by Tolman and Ray has shown that, from microscopic evidence, we may believe that bornite gives off some of its iron and becomes chalcocite.

With a view of actually securing some data on this idea samples of bornite were analysed for their iron contents. These powdered samples were placed in water in sealed tubes and were heated in an autoclave at a temperature of about 175 degrees for two hours.

The contents of the tubes were filtered and the residues were washed with N/10 sulfuric acid as this acid does not attack bornite to any appreciable extent.

The filtrates and washings gave strong iron tests.

Quantitative tests were not made.

Experiment No. 43.—The dispersing power of hydrogen sulfide in acid solution.

To determine the dispersing power (relative), in acid

solution four gas washing bottles were taken. To the first 2 grams of 200 mesh chalcocite was added, to the second 2 grams of 200 mesh covellite, to the third bornite and to the fourth chalcopyrite. 50 cc of N/10 sulfuric acid was added to each bottle. H_2S was passed thru the 4 bottles for a few minutes each day for 4 weeks.

At the end of this time all the minerals were slightly dispersed, this dispersion appearing most in the chalcocite and least in the chalcopyrite.

The contents of the bottles were filtered. All of the filtrates appeared clear, tho that from the chalcocite gave the merest suggestion of being less clear than the others.

Lumps of zinc were added to each filtrate. In the chalcocite filtrate a brownish black cloud appeared at once around the zinc. This cloud did not appear in the others tho they became less clear.

The residues which were left in the tubes containing the filtrates after the zinc had gone into solution were dissolved in nitric acid and this acid was more than neutralized with NH_4OH. All showed traces of copper the test being strongest in the residue from the chalcocite filtrate and least in that from the chalcopyrite. A very noticeable amount of $Fe(OH)_3$ was observed in the tube containing the residue from the bornite filtrate.

This experiment showed clearly that H_2S disperses these minerals in acid solution and that the dispersion is in a measure proportional to their copper content. It also shows that in acid solution and in contact with H_2S bornite gives up some of its iron.

Dispersion in acid solution is very much less then in alkaline.

Experiment No. 44.—Order of dispersion of minerals in alkaline solution.

Under conditions as described in experiment No. 43 chalcopyrite, bornite, enargite, covellite and chalcocite were covered with dilute KHS and were treated with H_4S for 4 days. All dispersed. An attempt was made to show that the iron to copper ratio in the dispersed material was less than in the mineral from which the material was dispersed. Because of limited time colorimetric methods were used. These were not suitable because of the iron content of even the best reagents. This experiment did show that these minerals were dispersed in the reverse order as given, and dispersed in such a finely divided form that the material readily went thru a filter. This dispersion was very strong.

SUMMARY.

While the work included in this present paper has been largely exploration along a variety of lines, and while it can scarcely be said that the investigation has been carried in any particular direction to such an extent as to ultimately determine with quantitative accuracy the importance of the various factors studied in respect to the enrichment of sulfide ores, it seems nevertheless true that certain heretofore little understood and in many cases unsuggested processes do enter into the general problem of the chemistry of the enrichment of the sulfide ores of copper. Perhaps the whole results may be summarized in the following general statements:

(1) Enrichment does not necessarily depend upon oxidation and leaching processes and upon the formation of electrolytic solutions of copper, such as the sulfate or chloride, but may perfectly well occur thru the intervention of the colloidal dispersion and solution, and subsequent deposition, either in the amorphous or crystalline form of already existing sulfides of copper. This applies to *upward* enrichment.

(2) The most effective dispersing agent which is likely to be found in nature is hydrogen sulfide, which may be produced in a variety of ways, notably from any sulfide minerals by contact with dilute acids or even with pure water. It is probably assisted by carbon dioxide.

The spontaneous dispersion of already existing sulfides is least in acid solutions and enormously greater in alkaline solution, from which it is to be expected that such operations will be of greater importance at considerable depth.

Free sulfur may also serve as a dispersing agent, and the same is true of higher sulfur-bearing minerals as pyrite, which acts to all intents and purposes as potential sulfur.

(3) The conditions favorable to the deposition of dispersed copper sulfides are the removal or absorption of the dispersing agent. Thus loss of hydrogen sulfide from a colloidal suspension of sulfides of copper

tends to flocculate and precipitate these sulfides. The condition for the appearance of crystalline rather than amorphus deposits seems to be the fairly complete removal of hydrogen sulfide which is readily accomplished by sulfur dioxide which accounts for the success of the Winchell-Tolman experiments.

(4) There seems to be no doubt but that cuprous sulfide (chalcocite) is the most stable of all of the copper sulfide minerals, in contact with solution, at least, and that there will be a general tendency for all copper sulfide-bearing minerals to eliminate cuprous sulfide, and for all sulfide precipitates from electrolyte solutions to spontaneously go over into the cuprous form. This tendency may possibly be more or less hampered by the presence of an excess of sulfur, such sulfur may tho be gradually eliminated by solution in alkaline waters, and even by reaction with pure water, and transportation to long distances may be accomplished.

Thus in the following figure, (diagramatic), if AB represents a mass of sulfide-bearing ore, and alkaline waters charged with hydrogen sulfide gradually percolate from B toward A, there will occur first a zone of dispersion (Zone 2). Upon elimination or absorption of hydrogen sulfide, there will follow a zone of amorphous chalcocite enrichment (Zone 3).

Following this with more complete elimination of hydrogen sulfide, accomplished perhaps by the appearance of sulfur dioxide as a result of limited oxygen supply, crystalline chalcocite will appear, (Zone 4).

A		
Appearance of SO_2 thru small supply of O_2. Deposit of crystalline chalcocite.	Zone 4	
Elimination or absorption of H_2S. Deposit of amorphous chalcocite changing to massive.	Zone 3	Copper
		Sulfide
Zone of dispersion.	Zone 2	Bearing
		Zones.
Alkaline water charged with H_2S.	Zone 1	
B		

(5) There seems to be also a curious physical effinity on the part of the chalcocite to draw toward itself any freshly formed cuprous sulfide, and to cause amorphous deposits of the new sulfide on the already present chalcocite. Already present chalcocite may also seemingly exert a sort of induced acceleration on the reaction between electrolytic solution of copper salts and free sulfur, resulting in the ultimate formation of chalcocite. (See Experiment No. 13.)

APPENDIX.

After the author had finished the experimentation which led to the conclusions given in this paper, there remained many questions bearing on this work which still needed explanation. Time was not available for anything like a complete investigation of these questions but the author was able to make some interesting experiments which are suggestive. Since the conclusions as derived from these experiments have only a suggestive value the conclusions and the experiments are here given in this appendix.

In this paper it has been shown that copper sulfide is kept colloidal by hydrogen sulfide and with hydrogen sulfide the colloid, (or colloids), migrate very readily. Naturally the question arose as to whether this dispersion and migration would be assisted or retarded by carbon dioxide, which is known to be present in many of the springs which give off hydrogen sulfide, and it is found in some mine gases which are given off from the cracks in the wall rocks of the mines. Quite naturally, too, it was asked whether some solutions would favor dispersion and whether others would tend to flocculate the colloidal copper sulfide. What effect would certain wall substances have on the colloid? Moreover it has been asked very frequently, what becomes of sulfur if such at any time be set free in the process of ore deposition.

One sees in experiments No. 45, No. 46, and No. 47 some interesting facts which suggest possible answers to these questions.

We notice here that those solutions containing small

amounts of sodium and potassium salts are most favorable for the dispersion of the colloidal copper sulfide, and believe that our deep-seated ore solutions actually are rich in potassium. We see that calcium and aluminium salts coagulate the colloidal copper sulfide and notice particularly that even the most insoluble aluminium compound, (dehydrated aluminium oxide), is exceedingly effective as a flocculant. We notice that calcium carbonate flocculates the colloid but that in the case of this colloidal copper sulfide being precipitated by calcium carbonate the colloid may be slightly dispersed again by means of hydrogen sulfide, some calcium going into solution, altho this dispersion is by no means as great as in the presence of alkali salts.

If one makes but a hasty examination of a good treatise on ore deposits* he will notice many examples of copper sulfide deposits embedded in limestone, some of the limestone having been replaced, and he will find that equally numerous are the data showing copper sulfide deposits in contact with, and embedded in, shale, gouge or other argillaceous material. May not the results of these experiments partially account for such deposits?

As is shown in Experiment No. 46, carbon dioxide is a most excellent agent for dispersing colloidal sulfur,

---

*—Very strikingly shown in many illustrations in Lindgren, "Mineral Deposits".

tho, as the author has found, it assists the dispersion of copper sulfide when hydrogen sulfide is present and precipitates the copper sulfide when hydrogen sulfide is absent. We may not think that carbon dioxide has aided in bringing the copper sulfide up to where it is deposited, has caused the colloidal sulfide to flocculate when the hydrogen sulfide has escaped, has been assimilated or has ceased to flow, and then having caused the copper sulfide to deposit, the carbon dioxide has carried on any excess sulfur to be used up in pyritization of the surrounding country rock and perhaps in many other ways?

More thorogoing investigation of all the lines suggested in this paper will be pushed forward as rapidly as possible.

Experiment No. 45.—Flocculation of colloidal copper sulfide.

Some colloidal copper sulfide was prepared, and by washing was carefully freed from the presence of any electrolyte.

Equal portions of this colloidal material were put into flasks and were treated with dilute solutions of NaCl, KCl, $CaCl_2$ and $AlCl_3$ and with powdered $CaCO_3$ and Al_2O_3 in water as can be seen in detail in the accompanying table.

It was found that the presence of NaCl and KCl increased the amount of the sulfide which was held in suspension, whereas $CaCl_2$, $CaCO_3$, $AlCl_3$, and Al_2O_3 caused the colloid to flocculate. This flocculation was accomplished very rapidly by the Al_2O_3.

were treated with H_2S. The hydrogen sulfide increased the amount of material which was held in suspension by water, NaCl solution and KCl solution. Also it carried into suspension some copper sulfide which had been flocculated at $CaCO_3$, thus indicating that tho limestone may tend to precipitate copper sulfide from ore solutions having this in suspension, the colloids in the presence of H_2S may, to some extent at least, migrate thru limestone. The H_2S failed to disperse the colloid which had been precipitated by $CaCl_2$, $AlCl_3$. and Al_2O_3.

A similar set of solutions with the colloidal copper sulfide were treated with a mixture of H_2S and CO_2. The results were much as those produced with the H_2S alone, except that not as much of the colloid, which had been precipitated by the $CaCO_3$ was again put into suspension, which could be accounted for by the fact that the CO_2 acting on the carbonate undoubtedly produced calcium ions.

This experiment showed most strikingly the great tendency of argillaceous material to flocculate colloid copper sulfide.

	1 Colloidal CuS in water	2 Colloidal CuS in N/300 NaCl	3 Colloidal CuS in N/300 KCl	4 Colloidal CuS in N/300 CaCl₂	5 Colloidal CuS with 1 gram CaCO₃	6 Colloidal CuS in N/300 AlCl₃	7 Colloidal CuS with 1 gram Al₂O₃
Suspended colloid	Suspended colloid	Remained suspended — More in suspension than in 1 A	More in suspension than in 1 A	Colloid soon flocculated	Colloid soon flocculated	Colloid soon flocculated	Colloid flocculated at once
Liquid satrated with H₂S	Suspended colloid — H₂S caused the amount of suspended colloid to increase	H₂S caused the amount of suspended colloid to increase	H₂S caused the amount of suspended colloid to increase very much	H₂S failed to cause the flocculated colloid to disperse	H₂S dispersed the flocculated colloid to some extent	H₂S failed to cause the flocculated colloid to disperse	H₂S failed to cause the flocculated colloid to disperse
Liquid saturated with H₂S and CO₂	Suspended colloid — H₂S and CO² caused more colloid to be suspended than in 1 B	H₂S and CO² caused more colloid to be suspended than in 2 B	H₂S and CO² caused more colloid to be suspended than in 3 B	H₂S and CO² failed to cause the flocculated colloid to disperse	H₂S and CO² failed to cause the flocculated colloid to disperse as much as in 5 B	H₂S and CO² failed to cause the flocculated colloid to disperse	H₂S and CO² failed to cause the flocculated colloid to disperse

Experiment No. 46.—Flocculation of colloidal sulfur.

Solutions of the substances in Experiment No. 45 were added to colloidal sulfur, and then tions containing the colloid were treated with H_2S and with the H_2S and CO_2 mixture the material was treated in Experiment No. 45.

The results are here tabulated.

	1 Colloidal S in water	2 Colloidal S in N/300 NaCl	3 Colloidal S in N/300 KCl	4 Colloidal S in N/300 $CaCl_2$	5 Colloidal S with 1 gram $CaCO_3$	6 Colloidal S N/300 $AlCl_3$
A Suspended colloid	S remained suspended	More S suspended than in 1 A	More S suspended than in 1 A	S soon flocculated	S soon flocculated	S soon flocculated
B Suspended colloid Liquid saturated with H_2S	S flocculated rapidly	S flocculated slowly	S flocculated slowly	S remained flocculated	S remained flocculated	S remained flocculated
C Suspended colloid Liquid saturated with H_2S and CO_2	Much S remained suspended tho not as much as in 1 A	Much S remained suspended tho not as much as in 2 A	Much S remained suspended tho not as much as in 3 A	S dispersed to some slight extent	S dispersed to some slight extent	S remained flocculated

Experiment No. 47.—Dispersing and flocculating effects of H_2S, SO_2, and CO_2 on colloidal copper sulfide and on colloidal sulfur, in water and in the presence of $N/1400$ K_2CO_3 and KCl.

A colloidal copper sulfide was prepared as described in Experiment No. 45. A colloidal sulfur solution was prepared by the action of H_2S on SO_2, the sulfur being suspended in water.

Portions of each colloidal solution were saturated with H_2S, SO_2, and CO_2, and then portions to which K_2CO_3 and KCl had been added so that the K ion was $N/1400$, were saturated with the gases.

The results are here tabulated.

SULFUR

	Water suspension of the Colloid	$N/1400$ K_2CO_3 and KCl suspension of the Colloid.
H_2S	H_2S caused the sulfur to flocculate rapidly.	H_2S caused the sulfur to flocculate slowly.
SO_2	SO_2 did not flocculate the sulfur rapidly.	SO_2 did not flocculate the sulfur rapidly.
CO_2	CO_2 practically prevented the flocculation of the sulfur.	CO_2 prevented the flocculation of the sulfur.

COPPER SULFIDE

	Water suspension of the Colloid.	$N/1400$ K_2CO_3 and KCl suspension of the Colloid.
H_2S	H_2S kept the copper sulfide dispersed for a long time.	H_2S practically prevented any flocculation of the copper sulfide.
SO_2	SO_2 caused the copper sulfide to flocculate.	SO_2 caused the copper sulfide to flocculate.
CO_2	CO_2 caused a very clean cut flocculation of the copper sulfide.	CO_2 caused a very clean cut flocculation of the colloid.

BIBLIOGRAPHY.*

Austin, L. S. "The metallurgy of the common metals" 2d edit. p 281.

Bain, H. F. Sedi-genetic and igneo-genetic ores" Econ. Geol. *1*, pp 331-339, 1906.

Bastin, E. S. "Metasomatism in downward sulfide enrichment" Econ. Geol. *8*, pp 51-63, 1913.

Beck, R. "The nature of ore deposits" trans by W. H. Weed, pp 361-382.

Berhmer, M. "The localization of values in ore bodies, and the occurrence of shoots in metalliferous deposits: secondary enrichment and impoverishment" Econ. Geol. *3*, pp 337-340, 1908.

Boutwell, J. M. "Economic geology of the Bingham Mining District, Utah." U. S. G. S. prof. paper 38, pp 212-217, pp 218-223, pp 228-235, and photographs of primary and secondary ores, plates XIV to XXXVIII.

Brokaw, A. D. "The precipitation of gold by manganese salts" Jour. Ind. & Eng. Chem., July, 1913. "The secondary precipitation of gold in ore bodies" Jour. Geol. *21*, pp 251-268, 1913.

Buehler and Gottschalk, "The oxidation of sulfides" Econ. Geol. *5*, pp 28-37, 1910; *7* pp 15-35, 1912.

Clifford, J. O. "Formation and growth of disseminated copper deposits" Mines and Methods, June, 1913.

Cooke, H. C. "The secondary enrichment of silver ores" Jour. Geol. *21*, pp 1-29, 1913.

---

*—This bibliography covers the most important publication on enrichment. It is particularly fertile in references to **downward** enrichment. References used by the author are given in the text of this article.

Doelter, C. "Ueber die Umwandlung amorpher Korper in Kristallinische." Zeit. f. Chemie u. Industrie der Kolloide, *3*, No. 1, p 29; No. 2, p 86.

Dutoit, P. "Sur les hyposulfites cupro-alkalins". Jour. de Chemie Physique, *2*, No. 4, pp 650-673, 1913.

Editorials. Mining and Scientific Press, *107*, 920. Engineering and Mining Journal, *95*, 463.

Elsing, M. T. "Relations of outcrops to ore at Cananea." Eng. & Min. Jour., *95*, pp 357-369, 1913.

Emmens, S. H. "The chemistry of the gossan." Eng. & Min. Jour., *44*, pp 582-583, 1892.

Emmons, S. F. "The secondary enrichment of ore deposits." Trans. A. I. M. E., *30*, pp 177-217.

Emmons, S. F., and Tower, G. W. "Economic geology of the Butte Special District, Mont." Folio 48, U. S. G. S., 1901.

Emmons. W. H. "The Cashin Mine, Montrose County, Colo." U. S. G. S. Bull. 285, pp 125-128, 1906.

Emmons, W. H. "The Enrichment of Sulfide Ores." U. S. G. S. Bull., 529, 1913.

This is the most important and exhaustive treatment of the subject of enrichment which has yet appeared. Page references are far too numerous to be listed.

Emmons, W. H. "A genetic classification of minerals." Econ. Geol. *3*, pp 611-627, 1908.

Emmons, W. H. "Secondary enrichment in Granite-Bimetallic Mine, Phillipsburg, Mont." Science, n. s., *27*, p 925, 1908.

Emmons, W. H. "Outcrops of ore bodies." Min. & Sci. Press, *99*, pp 751-759; pp 782-787, 1909; *100*, pp 162, 163, 1910.

Emmons, W. H. "Agency of manganese in the superficial alteration and secondary enrichment of gold deposits in the United States." Trans. A. I. M. E. Bull. 46, pp 767-837, 1910.

Farrel, J. H. "Practical Field Geology," 1912, Chapter XI.

Foote, H. W. "Criteria of downward sulfide enrichment." Econ. Geol. 5, pp 485-488, 1910.

Fuchs, E. and De Launay, L. "Traité des Gites Miniére et Métallifére," 2, pp 230-234, 1893.

Gmelin-Kraut. "Handbuch der anorganische Chemie," 5, p 807.

Graton, L. C. "The ore deposits of New Mexico." U. S. G. S. Prof. Paper No. 68, p 316, p 57, 1910.

Graton, L. C. "Investigation of copper enrichment." Eng. & Min. Jour., 96, pp 886-887.

Graton and Murdock. "The sulfide ores of copper, some results of microscopic study." Bull. A. I. M. E. 77, pp 741-786, 1913.

Gregory, J. W. "Criteria of downward sulfide enrichment." Econ. Geol., 5, pp 678-681, 1910.

Grout, F. F. "On the behavior of cold acid sulfate solutions of copper, silver, and gold, with alkaline extracts of metallic sulfides." Econ. Geol., 8, pp 407-433, 1913.

Gunther, C. G. "Examinations of Prospects," 1912, Chap. X.

Hillebrande, W. F. See U. S. G. S. Bull. 529, p 74.

Kemp, J. F. "Secondary enrichment in ore deposits of copper." Econ. Geol. 1, pp 11-25, 1905.

Kemp, J. F. "Some new points in the geology of copper ores." Min. & Sci. Press, 94, pp 402-403, 1907.

Kemp, J. F. "Influences of depth on metalliferous deposits", Min. & Eng. World, XXXIX, pp 394-591, 1913.

Keyes, C. R. "Genesis of the Lake Valley silver deposits." Trans. A. I. M. E. *39*, pp 131-169, 1909.

Keyes, C. R. "Criteria of downward sulfide enrichment." Econ. Geol. *5*, pp 558-564, 1910.

Keyes, C. R. "Porphry coppers." Bull. 25, Min. and Metallurgical Society, pp 316-320. 1910.

Kirk, C. R. "Conditions of mineralization in the copper veins at Butte, Mont." Econ. Geol., *7*, pp 35-83, 1912.

Krusch, P. "Primary and secondary ores considered with especial reference to the gel and the rich heavy metal ores", Min. & Sci. Press, *107*, pp 418-423, 1913.

Lakes, A. "Vulcan and Mammoth Chimney Mines." Trans A. I. M. E., *26*, pp 440-448. 1896.

Lane, A. C. Mich. Geol. & Biol. Sur. Pub. 6, Geol. Ser. 4, Vol. II.

Lazerovic, M "Einige Beitrage zu den Kriterien der reicher Sulfidzone." Zeit. f. prak. Geol., *19*, pp 321-467, 1911.

Lazerovic, M. "Die Enargit-Covellin Lakerstatte von Cuka-Dulkan bie Bor in Osterbien." Zeit. f. prak. Geol. **20**, pp 337-371, 1912.

Lepsius, B. "Ueber die Abnahme de gelosten Sauerstoffs in Grundwasser und einen einfachen Apparat zur Entanahme von Tiefproben in Bohrlochern." Ber. Deutsch. Chem. Ges., *18*, pt. 2, pp 2487-2490, 1885.

Linder and Picton. "Some physical properties of arsenious sulfide and other solutions." Jour. Chem. Soc., *67*, p 63, 1895.

Lindgren, W. "Mineral Deposits." Chap. XXIX, 1913.

Lindgren, W. "The genesis of the copper deposits of Clifton-Morenci." Trans. A. I. M. E. *35*, pp 511-550, 1905.

Lindgren, W. "Chemistry of copper deposits." Eng. & Min. Jour., *79*, p 189, 1905.

Lindgren, W. "The relation of ore deposition to physical conditions." Econ. Geol., *2*, pp 105-127, 1907.

Louis, Henry. "Criteria of downward sulfide enrichment." Econ. Geol., *5*, pp 81-89, 1910.

Lovemen, H. M. "Geology of the Miami Copper Mine." Min. & Sci. Press, *105*, pp 146-148, 1912.

Nishihara, G. "Sulfide enrichment." Min. & Sci. Press, Feb. 28, 1914.

Palmer and Bastin. "Metallic minerals as precipitants of silver and gold", Econ. Geol., *8*, pp 140-171, 1913.

Purington, C. W. "Secondary enrichment." Eng. & Min. Jour., *75*, pp 472, 473, 1903.

Ransome, F. L. "Geology of the Globe Copper District, Ariz." U. S. G. S. Prof. paper No. 12, pp 128-132, 1903.

Ransome, F. L. "Geology and ore deposits of the Bisbee Quadrangle." U. S. G. S. Prof. paper No. 21, pp 132-160, 1904.

Ransome, F. L. "Criteria of downward sulfide enrichment." Econ. Geol., *5*, pp 202-220, 1910.

Ransome, F. L. "Note on nomenclature of secondary ores." Econ. Geol., *8*, p 721, 1913.

Ray, I. C. (See Tolman).

Read, T. T. "The secondary enrichment of copper iron sulfides." Trans. A. I. M. E., *37*, pp 297-303, 1907.

Ries, H. "Economic Geology of the United States." 2d ed., pp 334-340.

Rogers, A. F. "Upward secondary sulfide enrichment and chalcocite formation at Butte, Mont." Econ. Geol., *8*, pp 781-794, 1913.

Sales, R. H. "Superficial alteration of the Butte veins." Econ. Geol., *5*, pp 15-21, 1910.

Sales, R. H. "Criteria of downward sulfide enrichment." Econ. Geol., *5*, pp 681, 682, 1910.

Sales, R. H. "Ore deposits at Butte, Mont." Bull. A. I. M. E., *80*, pp 1523-1627, 1913.

Schuermann, E. "Ueber die Verwandschaft der Schwermetalle zum schwefel." Liebig's Ann der Chemie, *249*, p 326, 1888.

Smith, G. "The secondary enrichment of ore deposits." Trans. A. I. M. E., *33*, pp 1055-1059, 1903.

Shannon, E. V. "Secondary enrichment in the Caledonia Mine, Coeur d'Alene, Idaho." Econ. Geol., *8*, pp 565-571, 1913.

Spencer, A. C. "Chalcocite deposition." Jour. Wash. Acad. Sci., *3*, pp 70-75, 1913.

Spencer, A. C. "Chalcocite enrichment." Econ. Geol., *8*, pp 621-652, 1913.

Spurr, J. E. "Geology as applied to mining," pp 265-293, 1904.

Spurr, J. E., and Garrey, G. H. "Economic geology of the Georgetown quadrangle." U. S. G. S. Prof. paper No. 63, 143, 1908.

Starbird, H. B. "Secondary enrichment in arid regions." Eng. & Min. Jour., *75*, pp 702, and 703, 1903.

Stokes, H. N. "On pyrite and marcasite." U. S. G. S. Bull. 186, pp 1-50, 1901.

Stokes, H. N. "Experiments on the solution, transportation and deposition of copper, silver and gold." Econ. Geol., *1*, pp 644-651, 1906.

Stokes, H. N. "Experiments on the action of various solutions on pyrite, and marcasite." Econ. Geol., *2*, pp 14-23, 1907.

Sullivan, E. C. "The secondary enrichment of copper iron sulfides." Trans. A. I. M. E., *13*, pp 143-145, 1907.

Sullivan, E. C. "The interaction between minerals and water solutions with special reference to geologic phenomena." U. S. G. S. Bull. *312*, 1907.

Tarr, W. A. "Copper in the 'Red Beds' of Oklahoma." Econ. Geol. *5*, pp 221-226, 1910.

Thompson, A. P. "The relations of pyrrhotite to chalcopyrite and other sulfides." Sch. Min. Quarterly, *XXXIV*, pp 385-395, 1913.

Tolman, C. F. Jr. "The southern Arizona copper fields." Min. & Sci. Press, *99*, 356-369, 390-393, 1909.

Tolman, C. F. Jr. "Disseminated chalcocite deposits at Ray, Ariz." Min. & Sci. Press, *99*, pp 622-624. 1909.

Tolman, C. F. Jr. "The Miami-Inspiration ore zone." Min. & Sci. Press *99*, 646-648, 1909.

Tolman, C. F. Jr. "Secondary Sulfide enrichment." Min. & Sci. Press, *106*, 38-43, 141-145, 178-181, 1913.

Tolman, J. C. Jr. and Ray, J. C. Unpublished manuscript.

Vogt, J. H. L. "Problems in the geology of ore deposits," in Posepny, Franz; "The genesis of ore deposits," p 675. 1902.

Wassjuchnowa, Fraulein. See reference. Zeitschr. f. Electrochemie, *10*. No. 22. p 902. 1913.

Weed, W. H. "The enrichment of gold and silver veins." Trans. A. I. M. E., *30*, pp 426-448. 1901.

Weed, W. H. "Ore deposits at Butte, Mont." U. S. G. S. Bull., 213, pp 170-180, 1903.

Weed, W. H. "Enrichment of mineral veins by lower metallic sulfides." Bull. Geol. Soc. Am. *15*, pp 179-206, 1900.

Wells, R. C. "The fractional precipitation of sulfides." Econ. Geol., *5*, pp 1-19, 1910.

Wells, R. C. "Electro-chemical activity between solutions and ores." Econ. Geol., *8*, pp 571-578, 1913.

Wells, R. C. "The criteria of downward sulfide enrichment." Econ. Geol. *5*, 479-489, 1910.

Weigel, O. "Die loslichkeit von Schwermetallsulfiden in reinem Wasser." Zeit. f. phys. Chem., *58*, pp 293-300, 1907.

Whitman, A. R. "Vadose synthesis of pyrite." Econ. Geol., *8*, pp 455-467, 1913.

Winchell, A. N. "Criteria of downward sulfide enrichment." Econ. Geol. *5*, pp 488-491, 1910.

Winchell, H. V. "Synthesis of chalcocite and its genesis at Butte." Bull. Geol. Soc. Am. *14*, pp 269-276; and Eng. & Min. Jour. *75*, pp 782-784, 1903.

Winchell, H. V. "Persistence of ore in depth." Min. & Sci. Press, 107, pp 332-334, 1913.

This work was undertaken in the hope of giving more information to the geologist, and in some degree it has been successful.

The amount of success which has attended the efforts of the writer is in no small degree due to the interested suggestion, assistance and oversight given by

Young. Professor Tolman indicated geological interpretations of the work, Professor Rogers contributed most willingly many of his finest mineral specimens for experimentation and Professor Young gave direct supervision of the research. To these gentlemen, and particularly to Professor Young, the author acknowledges his deep indebtedness.

JOHN DUSTIN CLARK.

Leland Stanford Junior University,
        May 18th, 1914.

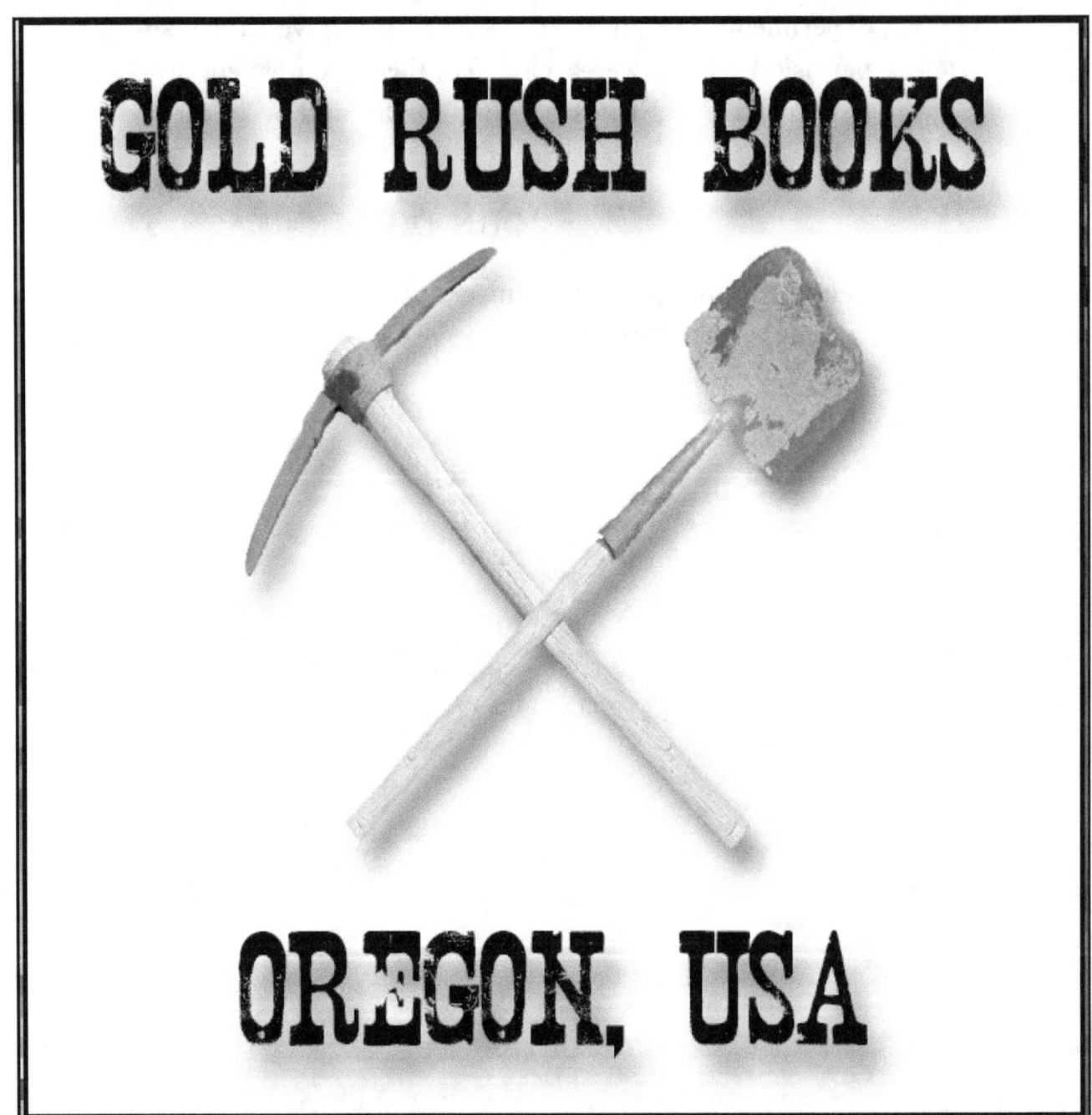

GOLD RUSH BOOKS

OREGON, USA

www.GoldMiningBooks.com

Books On Mining

Visit: www.goldminingbooks.com to order your copies or ask your favorite book seller to offer them.

Mining Books by Kerby Jackson

Gold Dust: Stories From Oregon's Mining Years - Oregon mining historian and prospector, Kerby Jackson, brings you a treasure trove of seventeen stories on Southern Oregon's rich history of gold prospecting, the prospectors and their discoveries, and the breathtaking areas they settled in and made homes. 5" X 8", 98 ppgs. Retail Price: $11.99

The Golden Trail: More Stories From Oregon's Mining Years - In his follow-up to "Gold Dust: Stories of Oregon's Mining Years", this time around, Jackson brings us twelve tales from Oregon's Gold Rush, including the story about the first gold strike on Canyon Creek in Grant County, about the old timers who found gold by the pail full at the Victor Mine near Galice, how Iradel Bray discovered a rich ledge of gold on the Coquille River during the height of the Rogue River War, a tale of two elderly miners on the hunt for a lost mine in the Cascade Mountains, details about the discovery of the famous Armstrong Nugget and others. 5" X 8", 70 ppgs. Retail Price: $10.99

Oregon Mining Books

Geology and Mineral Resources of Josephine County, Oregon - Unavailable since the 1970's, this important publication was originally compiled by the Oregon Department of Geology and Mineral Industries and includes important details on the economic geology and mineral resources of this important mining area in South Western Oregon. Included are notes on the history, geology and development of important mines, as well as insights into the mining of gold, copper, nickel, limestone, chromium and other minerals found in large quantities in Josephine County, Oregon. 8.5" X 11", 54 ppgs. Retail Price: $9.99

Mines and Prospects of the Mount Reuben Mining District - Unavailable since 1947, this important publication was originally compiled by geologist Elton Youngberg of the Oregon Department of Geology and Mineral Industries and includes detailed descriptions, histories and the geology of the Mount Reuben Mining District in Josephine County, Oregon. Included are notes on the history, geology, development and assay statistics, as well as underground maps of all the major mines and prospects in the vicinity of this much neglected mining district. 8.5" X 11", 48 ppgs. Retail Price: $9.99

The Granite Mining District - Notes on the history, geology and development of important mines in the well known Granite Mining District which is located in Grant County, Oregon. Some of the mines discussed include the Ajax, Blue Ribbon, Buffalo, Continental, Cougar-Independence, Magnolia, New York, Standard and the Tillicum. Also included are many rare maps pertaining to the mines in the area. 8.5" X 11", 48 ppgs. Retail Price: $9.99

Ore Deposits of the Takilma and Waldo Mining Districts of Josephine County, Oregon - The Waldo and Takilma mining districts are most notable for the fact that the earliest large scale mining of placer gold and copper in Oregon took place in these two areas. Included are details about some of the earliest large gold mines in the state such as the Llano de Oro, High Gravel, Cameron, Platerica, Deep Gravel and others, as well as copper mines such as the famous Queen of Bronze mine, the Waldo, Lily and Cowboy mines. This volume also includes six maps and 20 original illustrations. 8.5" X 11", 74 ppgs. Retail Price: $9.99

Metal Mines of Douglas, Coos and Curry Counties, Oregon - Oregon mining historian Kerby Jackson introduces us to a classic work on Oregon's mining history in this important re-issue of Bulletin 14C Volume 1, otherwise known as the Douglas, Coos & Curry Counties, Oregon Metal Mines Handbook. Unavailable since 1940, this important publication was originally compiled by the Oregon Department of Geology and Mineral Industries includes detailed descriptions, histories and the geology of over 250 metallic mineral mines and prospects in this rugged area of South West Oregon. 8.5" X 11", 158 ppgs. Retail Price: $19.99

Metal Mines of Jackson County, Oregon - Unavailable since 1943, this important publication was originally compiled by the Oregon Department of Geology and Mineral Industries includes detailed descriptions, histories and the geology of over 450 metallic mineral mines and prospects in Jackson County, Oregon. Included are such famous gold mining areas as Gold Hill, Jacksonville, Sterling and the Upper Applegate. **8.5" X 11", 220 ppgs. Retail Price: $24.99**

Metal Mines of Josephine County, Oregon - Oregon mining historian Kerby Jackson introduces us to a classic work on Oregon's mining history in this important re-issue of Bulletin 14C, otherwise known as the Josephine County, Oregon Metal Mines Handbook. Unavailable since 1952, this important publication was originally compiled by the Oregon Department of Geology and Mineral Industries includes detailed descriptions, histories and the geology of over 500 metallic mineral mines and prospects in Josephine County, Oregon. **8.5" X 11", 250 ppgs. Retail Price: $24.99**

Metal Mines of North East Oregon - Oregon mining historian Kerby Jackson introduces us to a classic work on Oregon's mining history in this important re-issue of Bulletin 14A and 14B, otherwise known as the North East Oregon Metal Mines Handbook. Unavailable since 1941, this important publication was originally compiled by the Oregon Department of Geology and Mineral Industries and includes detailed descriptions, histories and the geology of over 750 metallic mineral mines and prospects in North Eastern Oregon. **8.5" X 11", 310 ppgs. Retail Price: $29.99**

Metal Mines of North West Oregon - Oregon mining historian Kerby Jackson introduces us to a classic work on Oregon's mining history in this important re-issue of Bulletin 14D, otherwise known as the North West Oregon Metal Mines Handbook. Unavailable since 1951, this important publication was originally compiled by the Oregon Department of Geology and Mineral Industries and includes detailed descriptions, histories and the geology of over 250 metallic mineral mines and prospects in North Western Oregon. **8.5" X 11", 182 ppgs. Retail Price: $19.99**

Mines and Prospects of Oregon - Mining historian Kerby Jackson introduces us to a classic mining work by the Oregon Bureau of Mines in this important re-issue of The Handbook of Mines and Prospects of Oregon. Unavailable since 1916, this publication includes important insights into hundreds of gold, silver, copper, coal, limestone and other mines that operated in the State of Oregon around the turn of the 19th Century. Included are not only geological details on early mines throughout Oregon, but also insights into their history, production, locations and in some cases, also included are rare maps of their underground workings. **8.5" X 11", 314 ppgs. Retail Price: $24.99**

Lode Gold of the Klamath Mountains of Northern California and South West Oregon
(See California Mining Books)

Mineral Resources of South West Oregon - Unavailable since 1914, this publication includes important insights into dozens of mines that once operated in South West Oregon, including the famous gold fields of Josephine and Jackson Counties, as well as the Coal Mines of Coos County. Included are not only geological details on early mines throughout South West Oregon, but also insights into their history, production and locations. **8.5" X 11", 154 ppgs. Retail Price: $11.99**

Chromite Mining in The Klamath Mountains of California and Oregon
(See California Mining Books)

Southern Oregon Mineral Wealth - Unavailable since 1904, this rare publication provides a unique snapshot into the mines that were operating in the area at the time. Included are not only geological details on early mines throughout South West Oregon, but also insights into their history, production and locations. Some of the mining areas include Grave Creek, Greenback, Wolf Creek, Jump Off Joe Creek, Granite Hill, Galice, Mount Reuben, Gold Hill, Galls Creek, Kane Creek, Sardine Creek, Birdseye Creek, Evans Creek, Foots Creek, Jacksonville, Ashland, the Applegate River, Waldo, Kerby and the Illinois River, Althouse and Sucker Creek, as well as insights into local copper mining and other topics. **8.5" X 11", 64 ppgs. Retail Price: $8.99**

Geology and Ore Deposits of the Takilma and Waldo Mining Districts - Unavailable since the 1933, this publication was originally compiled by the United States Geological Survey and includes details on gold and copper mining in the Takilma and Waldo Districts of Josephine County, Oregon. The Waldo and Takilma mining districts are most notable for the fact that the earliest large scale mining of placer gold and copper in Oregon took place in these two areas. Included in this report are details about some of the earliest large gold mines in the state such as the Llano de Oro, High Gravel, Cameron, Platerica, Deep Gravel and others, as well as copper mines such as the famous Queen of Bronze mine, the Waldo, Lily and Cowboy mines. In addition to geological examinations, insights are also provided into the production, day to day operations and early histories of these mines, as well as calculations of known mineral reserves in the area. This volume also includes six maps and 20 original illustrations. **8.5" X 11", 74 ppgs. Retail Price: $9.99**

Gold Mines of Oregon - Oregon mining historian Kerby Jackson introduces us to a classic work on Oregon's mining history in this important re-issue of Bulletin 61, otherwise known as "Gold and Silver In Oregon". Unavailable since 1968, this important publication was originally compiled by geologists Howard C. Brooks and Len Ramp of the Oregon Department of Geology and Mineral Industries and includes detailed descriptions, histories and the geology of over 450 gold mines Oregon. Included are notes on the history, geology and gold production statistics of all the major mining areas in Oregon including the Klamath Mountains, the Blue Mountains and the North Cascades. While gold is where you find it, as every miner knows, the path to success is to prospect for gold where it was previously found. **8.5" X 11", 344 ppgs. Retail Price: $24.99**

Mines and Mineral Resources of Curry County Oregon - Originally published in 1916, this important publication on Oregon Mining has not been available for nearly a century. Included are rare insights into the history, production and locations of dozens of gold mines in Curry County, Oregon, as well as detailed information on important Oregon mining districts in that area such as those at Agness, Bald Face Creek, Mule Creek, Boulder Creek, China Diggings, Collier Creek, Elk River, Gold Beach, Rock Creek, Sixes River and elsewhere. Particular attention is especially paid to the famous beach gold deposits of this portion of the Oregon Coast. **8.5" X 11", 140 ppgs. Retail Price: $11.99**

Chromite Mining in South West Oregon - Originally published in 1961, this important publication on Oregon Mining has not been available for nearly a century. Included are rare insights into the history, production and locations of nearly 300 chromite mines in South Western Oregon. **8.5" X 11", 184 ppgs. Retail Price: $14.99**

Mineral Resources of Douglas County Oregon - Originally published in 1972, this important publication on Oregon Mining has not been available for nearly forty years. Included are rare insights into the geology, history, production and locations of numerous gold mines and other mining properties in Douglas County, Oregon. **8.5" X 11", 124 ppgs. Retail Price: $11.99**

Mineral Resources of Coos County Oregon - Originally published in 1972, this important publication on Oregon Mining has not been available for nearly forty years. Included are rare insights into the geology, history, production and locations of numerous gold mines and other mining properties in Coos County, Oregon. **8.5" X 11", 100 ppgs. Retail Price: $11.99**

Mineral Resources of Lane County Oregon - Originally published in 1938, this important publication on Oregon Mining has not been available for nearly seventy five years. Included are extremely rare insights into the geology and mines of Lane County, Oregon, in particular in the Bohemia, Blue River, Oakridge, Black Butte and Winberry Mining Districts. **8.5" X 11", 82 ppgs. Retail Price: $9.99**

Mineral Resources of the Upper Chetco River of Oregon: Including the Kalmiopsis Wilderness - Originally published in 1975, this important publication on Oregon Mining has not been available for nearly forty years. Withdrawn under the 1872 Mining Act since 1984, real insight into the minerals resources and mines of the Upper Chetco River has long been unavailable due to the remoteness of the area. Despite this, the decades of battle between property owners and environmental extremists over the last private mining inholding in the area has continued to pique the interest of those interested in mining and other forms of natural resource use. Gold mining began in the area in the 1850's and has a rich history in this geographic area, even if the facts surrounding it are little known. Included are twenty two rare photographs, as well as insights into the Becca and Morning Mine, the Emmly Mine (also known as Emily Camp), the Frazier Mine, the Golden Dream or Higgins Mine, Hustis Mine, Peck Mine and others. **8.5" X 11", 64 ppgs. Retail Price: $8.99**

Gold Dredging in Oregon - Originally published in 1939, this important publication on Oregon Mining has not been available for nearly seventy five years. Included are extremely rare insights into the history and day to day operations of the dragline and bucketline gold dredges that once worked the placer gold fields of South West and North East Oregon in decades gone by. Also included are details into the areas that were worked by gold dredges in Josephine, Jackson, Baker and Grant counties, as well as the economic factors that impacted this mining method. This volume also offers a unique look into the values of river bottom land in relation to both farming and mining, in how farm lands were mined, re-soiled and reclamated after the dredges worked them. Featured are hard to find maps of the gold dredge fields, as well as rare photographs from a bygone era. **8.5" X 11", 86 ppgs. Retail Price: $8.99**

Quick Silver Mining in Oregon - Originally published in 1963, this important publication on Oregon Mining has not been available for over fifty years. This publication includes details into the history and production of Elemental Mercury or Quicksilver in the State of Oregon. **8.5" X 11", 238 ppgs. Retail Price: $15.99**

Mines of the Greenhorn Mining District of Grant County Oregon - Originally published in 1948, this important publication on Oregon Mining has not been available for over sixty five years. In this publication are rare insights into the mines of the famous Greenhorn Mining District of Grant County, Oregon, especially the famous Morning Mine. Also included are details on the Tempest, Tiger, Bi-Metallic, Windsor, Psyche, Big Johnny, Snow Creek, Banzette and Paramount Mines, as well as prospects in the vicinities in the famous mining areas of Mormon Basin, Vinegar Basin and Desolation Creek. Included are hard to find mine maps and dozens of rare photographs from the bygone era of Grant County's rich mining history. **8.5" X 11", 72 ppgs. Retail Price: $9.99**

Geology of the Wallowa Mountains of Oregon: Part I (Volume 1) - Originally published in 1938, this important publication on Oregon Mining has not been available for nearly seventy five years. Included are details on the geology of this unique portion of North Eastern Oregon. This is the first part of a two book series on the area. Accompanying the text are rare photographs and historic maps. **8.5" X 11", 92 ppgs. Retail Price: $9.99**

Geology of the Wallowa Mountains of Oregon: Part II (Volume 2) - Originally published in 1938, this important publication on Oregon Mining has not been available for nearly seventy five years. Included are details on the geology of this unique portion of North Eastern Oregon. This is the first part of a two book series on the area. Accompanying the text are rare photographs and historic maps. **8.5" X 11", 94 ppgs. Retail Price: $9.99**

Field Identification of Minerals For Oregon Prospectors - Originally published in 1940, this important publication on Oregon Mining has not been available for nearly seventy five years. Included in this volume is an easy system for testing and identifying a wide range of minerals that might be found by prospectors, geologists and rockhounds in the State of Oregon, as well as in other locales. Topics include how to put together your own field testing kit and how to conduct rudimentary tests in the field. This volume is written in a clear and concise way to make it useful even for beginners. **8.5" X 11", 158 ppgs. Retail Price: $14.99**

The Bohemia Mining District of Oregon - Originally published in 1900, this important publication on Oregon Mining has not been available for over a century. Included in this volume are important insights into the famous Bohemia Mining District of Oregon, including the histories and locations of important gold mines in the area such as the Ophir Mine, Clarence, Acturas, Peek-a-boo, White Swan, Combination Mine, the Musick Mine, The California, White Ghost, The Mystery, Wall Street, Vesuvius, Story, Lizzie Bullock, Delta, Elsie Dora, Golden Slipper, Broadway, Champion Mine, Knott, Noonday, Helena, White Wings, Riverside and others. Also included are notes on the nearby Blue River Mining District. **8.5" X 11", 58 ppgs. Retail Price: $9.99**

The Gold Fields of Eastern Oregon - Unavailable since 1900, this publication was originally compiled by the Baker City Chamber of Commerce Offering important insights into the gold mining history of Eastern Oregon, "The Gold Fields of Eastern Oregon" sheds a rare light on many of the gold mines that were operating at the turn of the 19th Century in Baker County and Grant County in North Eastern Oregon. Some of the areas featured include the Cable Cove District, Baisely-Elkhorn, Granite, Red Boy, Bonanza, Susanville, Sparta, Virtue, Vaughn, Sumpter, Burnt River, Rye Valley and other mining districts. Included is basic information on not only many gold mines that are well known to those interested in Eastern Oregon mining history, but also many mines and prospects which have been mostly lost to the passage of time. Accompanying are numerous rare photos **8.5" X 11", 78 ppgs. Retail Price: $10.99**

Gold Mining in Eastern Oregon - Originally published in 1938, this important publication on Oregon Mining has not been available for over a century. Included in this volume are important insights into the famous mining districts of Eastern Oregon during the late 1930's. Particular attention is given to those gold mines with milling and concentrating facilities in the Greenhorn, Red Boy, Alamo, Bonanza, Granite, Cable Cove, Cracker Creek, Virtue, Keating, Medical Springs, Sanger, Sparta, Chicken Creek, Mormon Basin, Connor Creek, Cornucopia and the Bull Run Mining Districts. Some of the mines featured include the Ben Harrison, North Pole-Columbia, Highland Maxwell, Baisley-Elkhorn, White Swan, Balm Creek, Twin Baby, Gem of Sparta, New Deal, Gleason, Gifford-Johnson, Cornucopia, Record, Bull Run, Orion and others. Of particular interest are the mill flow sheets and descriptions of milling operations of these mines. **8.5" X 11", 68 ppgs. Retail Price: $8.99**

The Gold Belt of the Blue Mountains of Oregon - Originally published in 1901, this important publication on Oregon Mining has not been available for over a century. Included in this volume are rare insights into the gold deposits of the Blue Mountains of North East Oregon, including the history of their early discovery and early production. Extensive details are offered on this important mining area's mineralogy and economic geology, as well as insights into nearby gold placers, silver deposits and copper deposits. Featured are the Elkhorn and Rock Creek mining districts, the Pocahontas district, Auburn and Minersville districts, Sumpter and Cracker Creek, Cable Cove, the Camp Carson district, Granite, Alamo, Greenhorn, Robinsonville, the Upper Burnt River Valley and Bonanza districts, Susanville, Quartzburg, Canyon Creek, Virtue, the Copper Butte district, the North Powder River, Sparta, Eagle Creek, Cornucopia, Pine Creek, Lower Powder River, the Upper Snake River Canyon, Rye Valley, Lower Burnt River Valley, Mormon Basin, the Malheur and Clarks Creek districts, Sutton Creek and others. Of particular interest are important details on numerous gold mines and prospects in these mining districts, including their locations, histories, geology and other important information, as well as information on silver, copper and fire opal deposits. **8.5" X 11", 250 ppgs. Retail Price: $24.99**

Mining in the Cascades Range of Oregon - Originally published in 1938, this important publication on Oregon Mining has not been available for over seventy five years. Included in this volume are rare insights into the gold mines and other types of metal mines in the Cascades Mountain Range of Oregon. Some of the important mining areas covered include the famous Bohemia Mining District, the North Santiam Mining District, Quartzville Mining District, Blue River Mining District, Fall Creek Mining District, Oakridge District, Zinc District, Buzzard-Al Sarena District, Grand Cove, Climax District and Barron Mining District. Of particular interest are important details on over 100 mines and prospects in these mining districts, including their locations, histories, geology and other important information. **8.5" X 11", 170 ppgs. Retail Price: $14.99**

Idaho Mining Books

Gold in Idaho - Unavailable since the 1940's, this publication was originally compiled by the Idaho Bureau of Mines and includes details on gold mining in Idaho. Included is not only raw data on gold production in Idaho, but also valuable insight into where gold may be found in Idaho, as well as practical information on the gold bearing rocks and other geological features that will assist those looking for placer and lode gold in the State of Idaho. This volume also includes thirteen gold maps that greatly enhance the practical usability of the information contained in this small book detailing where to find gold in Idaho. **8.5" X 11", 72 ppgs. Retail Price: $9.99**

Geology of the Couer D'Alene Mining District of Idaho - Unavailable since 1961, this publication was originally compiled by the Idaho Bureau of Mines and Geology and includes details on the mining of gold, silver and other minerals in the famous Coeur D'Alene Mining District in Northern Idaho. Included are details on the early history of the Coeur D'Alene Mining District, local tectonic settings, ore deposit features, information on the mineral belts of the Osburn Fault, as well as detailed information on the famous Bunker Hill Mine, the Dayrock Mine, Galena Mine, Lucky Friday Mine and the infamous Sunshine Mine. This volume also includes sixteen hard to find maps. **8.5" X 11", 70 ppgs. Retail Price: $9.99**

The Gold Camps and Silver Cities of Idaho - Originally published in 1963, this important publication on Idaho Mining has not been available for nearly fifty years. Included are rare insights into the history of Idaho's Gold Rush, as well as the mad craze for silver in the Idaho Panhandle. Documented in fine detail are the early mining excitements at Boise Basin, at South Boise, in the Owyhees, at Deadwood, Long Valley, Stanley Basin and Robinson Bar, at Atlanta, on the famous Boise River, Volcano, Little Smokey, Banner, Boise Ridge, Hailey, Leesburg, Lemhi, Pearl, at South Mountain, Shoup and Ulysses, Yellow Jacket and Loon Creek. The story follows with the appearance of Chinese miners at the new mining camps on the Snake River, Black Pine, Yankee Fork, Bay Horse, Clayton, Heath, Seven Devils, Gibbonsville, Vienna and Sawtooth City. Also included are special sections on the Idaho Lead and Silver mines of the late 1800's, as well as the mining discoveries of the early 1900's that paved the way for Idaho's modern mining and mineral industry. Lavishly illustrated with rare historic photos, this volume provides a one of a kind documentary into Idaho's mining history that is sure to be enjoyed by not only modern miners and prospectors who still scour the hills in search of nature's treasures, but also those enjoy history and tromping through overgrown ghost towns and long abandoned mining camps. **8.5" X 11", 186 ppgs. Retail Price: $14.99**

Ore Deposits and Mining in North Western Custer County Idaho - Unavailable since 1913, this important publication was originally published by the Us Department of the Interior and has been unavailable for a century. Included are fine details on the geology, geography, gold placers and gold and silver bearing quartz veins of the mining region of North West Custer County, Idaho. Of particular interest is a rare look at the mines and prospects of the region, including those such as the Ramshorn Mine, SkyLark, Riverview, Excelsior, Beardsley, Pacific, Hoosier, Silver Brick, Forest Rose and dozens of others in the Bay Horse Mining District. Also covered are the mines of the Yankee Fork District such as the Lucky Boy, Badger, Black, Enterprise, Charles Dickens, Morrison, Golden Sunbeam, Montana, Golden Gate and others, as well as those in the Loon Mining District. **8.5" X 11", 126 ppgs. Retail Price: $12.99**

Gold Rush To Idaho - Unavailable since 1963, this important publication was originally published by the Idaho Bureau of Mines and has been unavailable for 50 years. "Gold Rush To Idaho" revisits the earliest years of the discovery of gold in Idaho Territory and introduces us to the conditions that the pioneer gold seekers met when they blazed a trail through the wilderness of Idaho's mountains and discovered the precious yellow metal at Oro Fino and Pierce. Subsequent rushes followed at places like Elk City, Newsome, Clearwater Station, Florence, Warrens and elsewhere. Of particular interest is a rare look at the hardships that the first miners in Idaho met with during their day to day existences and their attempts to bring law and order to their mining camps. **8.5" X 11", 88 ppgs. Retail Price: $9.99**

The Geology and Mines of Northern Idaho and North Western Montana - Unavailable since 1909, this important publication was originally published by the Us Department of the Interior and has been unavailable for a century. Included are fine details on the geology and geography of the mining regions of Northern Idaho and North Western Montana. Of particular interest is a rare look at the mines and prospects of the region, including those in the Pine Creek Mining District, Lake Pend Oreille district, Troy Mining District, Sylvanite District, Cabinet Mining District, Prospect Mining District and the Missoula Valley. Some of the mines featured include the Iron Mountain, Silver Butte, Snowshoe, Grouse Mountain Mine and others. **8.5" X 11", 142 ppgs. Retail Price: $12.99**

Utah Mining Books

Fluorite in Utah - Unavailable since 1954, this publication was originally compiled by the USGS, State of Utah and U.S. Atomic Energy Commission and details the mining of fluorspar, also known as fluorite in the State of Utah. Included are details on the geology and history of fluorspar (fluorite) mining in Utah, including details on where this unique gem mineral may be found in the State of Utah. **8.5" X 11", 60 ppgs. Retail Price: $8.99**

California Mining Books

The Tertiary Gravels of the Sierra Nevada of California - Mining historian Kerby Jackson introduces us to a classic mining work by Waldemar Lindgren in this important re-issue of The Tertiary Gravels of the Sierra Nevada of California. Unavailable since 1911, this publication includes details on the gold bearing ancient river channels of the famous Sierra Nevada region of California. **8.5" X 11", 282 ppgs. Retail Price: $19.99**

The Mother Lode Mining Region of California - Unavailable since 1900, this publication includes details on the gold mines of California's famous Mother Lode gold mining area. Included are details on the geology, history and important gold mines of the region, as well as insights into historic mining methods, mine timbering, mining machinery, mining bell signals and other details on how these mines operated. Also included are insights into the gold mines of the California Mother Lode that were in operation during the first sixty years of California's mining history. **8.5" X 11", 176 ppgs. Retail Price: $14.99**

Lode Gold of the Klamath Mountains of Northern California and South West Oregon - Unavailable since 1971, this publication was originally compiled by Preston E. Hotz and includes details on the lode mining districts of Oregon and California's Klamath Mountains. Included are details on the geology, history and important lode mines of the French Gulch, Deadwood, Whiskeytown, Shasta, Redding, Muletown, South Fork, Old Diggings, Dog Creek (Delta), Bully Choop (Indian Creek), Harrison Gulch, Hayfork, Minersville, Trinity Center, Canyon Creek, East Fork, New River, Denny, Liberty (Black Bear), Cecilville, Callahan, Yreka, Fort Jones and Happy Camp mining districts in California, as well as the Ashland, Rogue River, Applegate, Illinois River, Takilma, Greenback, Galice, Silver Peak, Myrtle Creek and Mule Creek districts of South Western Oregon. Also included are insights into the mineralization and other characteristics of this important mining region. **8.5" X 11", 100 ppgs. Retail Price: $10.99**

Mines and Mineral Resources of Shasta County, Siskiyou County, Trinity County: California - Unavailable since 1915, this publication was originally compiled by the California State Mining Bureau and includes details on the gold mines of this area of Northern California. Also included are insights into the mineralization and other characteristics of this important mining region, as well as the location of historic gold mines. **8.5" X 11", 204 ppgs. Retail Price: $19.99**

Geology of the Yreka Quadrangle, Siskiyou County, California - Unavailable since 1977, this publication was originally compiled by Preston E. Hotz and includes details on the geology of the Yreka Quadrangle of Siskiyou County, California. Also included are insights into the mineralization and other characteristics of this important mining region. **8.5" X 11", 78 ppgs. Retail Price: $7.99**

Mines of San Diego and Imperial Counties, California - Originally published in 1914, this important publication on California Mining has not been available for a century. This publication includes important information on the early gold mines of San Diego and Imperial County, which were some of the first gold fields mined in California by early Spanish and Mexican miners before the 49ers came on the scene. Included are not only details on early mining methods in the area, production statistics and geological information, but also the location of the early gold mines that helped make California "The Golden State". Also included are details on the mining of other minerals such as silver, lead, zinc, manganese, tungsten, vanadium, asbestos, barite, borax, cement, clay, dolomite, fluospar, gem stones, graphite, marble, salines, petroleum, stronium, talc and others. **8.5" X 11", 116 ppgs. Retail Price: $12.99**

Mines of Sierra County, California - Unavailable since 1920, this publication was originally compiled by the California State Mining Bureau and includes details on the gold mines of Sierra County, California. Also included are insights into the mineralization and other characteristics of this important mining region, as well as the location of historic gold mines. **8.5" X 11", 156 ppgs. Retail Price: $19.99**

Mines of Plumas County, California - Unavailable since 1918, this publication was originally compiled by the California State Mining Bureau and includes details on the gold mines of Plumas County, California. Also included are insights into the mineralization and other characteristics of this important mining region, as well as the location of historic gold mines. 8.5" X 11", 200 ppgs. Retail Price: $19.99

Mines of El Dorado, Placer, Sacramento and Yuba Counties, California - Originally published in 1917, this important publication on California Mining has not been available for nearly a century. This publication includes important information on the early gold mines of El Dorado County, Placer County, Sacramento County and Yuba County, which were some of the first gold fields mined by the Forty-Niners during the California Gold Rush. Included are not only details on early mining methods in the area, production statistics and geological information, but also the location of the early gold mines that helped make California "The Golden State". Also included are insights into the early mining of chrome, copper and other minerals in this important mining area. 8.5" X 11", 204 ppgs. Retail Price: $19.99

Mines of Los Angeles, Orange and Riverside Counties, California - Originally published in 1917, this important publication on California Mining has not been available for nearly a century. This publication includes important information on the early gold mines of Los Angeles County, Orange County and Riverside County, which were some of the first gold fields mined in California by early Spanish and Mexican miners before the 49ers came on the scene. Included are not only details on early mining methods in the area, production statistics and geological information, but also the location of the early gold mines that helped make California "The Golden State". 8.5" X 11", 146 ppgs. Retail Price: $12.99

Mines of San Bernadino and Tulare Counties, California - Originally published in 1917, this important publication on California Mining has not been available for nearly a century. This publication includes important information on the early gold mines of San Bernadino and Tulare County, which were some of the first gold fields mined in California by early Spanish and Mexican miners before the 49ers came on the scene. Included are not only details on early mining methods in the area, production statistics and geological information, but also the location of the early gold mines that helped make California "The Golden State". Also included are details on the mining of other minerals such as copper, iron, lead, zinc, manganese, tungsten, vanadium, asbestos, barite, borax, cement, clay, dolomite, fluospar, gem stones, graphite, marble, salines, petroleum, stronium, talc and others. 8.5" X 11", 200 ppgs. Retail Price: $19.99

Chromite Mining in The Klamath Mountains of California and Oregon - Unavailable since 1919, this publication was originally compiled by J.S. Diller of the United States Department of Geological Survey and includes details on the chromite mines of this area of Northern California and Southern Oregon. Also included are insights into the mineralization and other characteristics of this important mining region, as well as the location of historic mines. Also included are insights into chromite mining in Eastern Oregon and Montana. 8.5" X 11", 98 ppgs. Retail Price: $9.99

Mines and Mining in Amador, Calaveras and Tuolumne Counties, California - Unavailable since 1915, this publication was originally compiled by William Tucker and includes details on the mines and mineral resources of this important California mining area. Included are details on the geology, history and important gold mines of the region, as well as insights into other local mineral resources such as asbestos, clay, copper, talc, limestone and others. Also included are insights into the mineralization and other characteristics of this important portion of California's Mother Lode mining region. 8.5" X 11", 198 ppgs. Retail Price: $14.99

The Cerro Gordo Mining District of Inyo County California - Unavailable since 1963, this publication was originally compiled by the United States Department of Interior. Included are insights into the mineralization and other characteristics of this important mining region of Southern California. Topics include the mining of gold and silver in this important mining district in Inyo County, California, including details on the history, production and locations of the Cerro Gordo Mine, the Morning Star Mine, Estelle Tunnel, Charles Lease Tunnel, Ignacio, Hart, Crosscut Tunnel, Sunset, Upper Newtown, Newtown, Ella, Perseverance, Newsboy, Belmont and other silver and gold mines in the Cerro Gordo Mining District. This volume also includes important insights into the fossil record, geologic formations, faults and other aspects of economic geology in this California mining district. 8.5" X 11", 104 ppgs. Retail Price: $10.99

Mining in Butte, Lassen, Modoc, Sutter and Tehama Counties of California - Unavailable since 1917, this publication was originally compiled by the United States Department of Interior. Included are insights into the mineralization and other characteristics of this important mining region of California. Topics include the mining of asbestos, chromite, gold, diamonds and manganese in Butte County, the mining of gold and copper in the Hayden Hill and Diamond Mountain mining districts of Lassen County, the mining of coal, salt, copper and gold in the High Grade and Winters mining districts of Modoc County, gold mining in Sutter County and the mining of gold, chromite, manganese and copper in Tehama County. This volume also includes the production records and locations of numerous mines in this important mining region. 8.5" X 11", 114 ppgs. Retail Price: $11.99

Mines of Trinity County California - Originally published in 1965, this important publication on California Mining has not been available for nearly fifty years. This publication includes important information on mines and mining in Trinity County, California, as well insights into the mineralization and geology of this important mining area in Northern California. Included are extensive details on hardrock and placer gold mines and prospects, including charts showing the locations of these historic mines.. **8.5" X 11", 144 ppgs. Retail Price: $12.99**

Mines of Kern County California - Originally published in 1962, this important publication on California Mining has not been available for nearly fifty years. This publication includes important information on mines and mining in Kern County, California, as well insights into the mineralization and geology of this important mining area in California. Included are extensive details on hardrock and placer gold mines and prospects, including charts showing the locations of these historic mines. **8.5" X 11", 398 ppgs. Retail Price: $24.99**

Mines of Calaveras County California - Originally published in 1962, this important publication on California Mining has not been available for nearly fifty years. This publication includes important information on mines and mining in Calaveras County, California, as well insights into the mineralization and geology of this important mining area in Northern California. Included are extensive details on hardrock and placer gold mines and prospects, including charts showing the locations of these historic mines. **8.5" X 11", 236 ppgs. Retail Price: $19.99**

Lode Gold Mining in Grass Valley California - Unavailable since 1940, this publication was originally compiled by the United States Department of Interior. Included are insights into the gold mineralization and other characteristics of this important mining region of Nevada County, California. This volume also includes important insights into the geologic formations, faults and other aspects of economic geology in this California mining district. Of particular interest are the fine details on many hardrock gold mines in the area, including their locations, histories, development and mineralization. Some of the mines featured include the Gold Hill Mine, Massachusetts Hill, Boundary, Peabody, Golden Center, North Star, Omaha, Lone Jack, Homeward Bound, Hartery, Wisconsin, Allison Ranch, Phoenix, Kate Hayes, W.Y.O.D., Empire, Rich Hill, Daisy Hill, Orleans, Sultana, Centennial, Conlin, Ben Franklin, Crown Point and many others. **8.5" X 11", 148 ppgs. Retail Price: $12.99**

Alaska Mining Books

Ore Deposits of the Willow Creek Mining District, Alaska - Unavailable since 1954, this hard to find publication includes valuable insights into the Willow Creek Mining District near Hatcher Pass in Alaska. The publication includes insights into the history, geology and locations of the well known mines in the area, including the Gold Cord, Independence, Fern, Mabel, Lonesome, Snowbird, Schroff-O'Neil, High Grade, Marion Twin, Thorpe, Webfoot, Kelly-Willow, Lane, Holland and others. **8.5" X 11", 96 ppgs. Retail Price: $9.99**

Arizona Mining Books

Mines and Mining in Northern Yuma County Arizona - Originally published in 1911, this important publication on Arizona Mining has not been available for over a hundred years. Included are rare insights into the gold, silver, copper and quicksilver mines of Yuma County, Arizona together with hard to find maps and photographs. Some of the mines and mining districts featured include the Planet Copper Mine, Mineral Hill, the Clara Consolidated Mine, Viati Mine, Copper Basin prospect, Bowman Mine, Quartz King, Billy Mack, Carnation, the Wardwell and Osbourne, Valensuella Copper, the Mariquita, Colonial Mine, the French American, the New York-Plomosa, Guadalupe, Lead Camp, Mudersbach Copper Camp, Yellow Bird, the Arizona Northern (Salome Strike), Bonanza (Harqua Hala), Golden Eagle, Hercules, Socorro and others. **8.5" X 11", 144 ppgs. Retail Price: $11.99**

The Aravaipa and Stanley Mining Districts of Graham County Arizona - Originally published in 1925, this important publication on Arizona Mining has not been available for nearly ninety years. Included are rare insights into the gold and silver mines of these two important mining districts, together with hard to find maps. **8.5" X 11", 140 ppgs. Retail Price: $11.99**

Gold in the Gold Basin and Lost Basin Mining Districts of Mohave County, Arizona - This volume contains rare insights into the geology and gold mineralization of the Gold Basin and Lost Basin Mining Districts of Mohave County, Arizona that will be of benefit to miners and prospectors. Also included is a significant body of information on the gold mines and prospects of this portion of Arizona. This volume is lavishly illustrated with rare photos and mining maps. **8.5" X 11", 188 ppgs. Retail Price: $19.99**

Mines of the Jerome and Bradshaw Mountains of Arizona - This important publication on Arizona Mining has not been available for ninety years. This volume contains rare insights into the geology and ore deposits of the Jerome and Bradshaw Mountains of Arizona that will be of benefit to miners and prospectors who work those areas. Included is a significant body of information on the mines and prospects of the Verde, Black Hills, Cherry Creek, Prescott, Walker, Groom Creek, Hassayampa, Bigbug, Turkey Creek, Agua Fria, Black Canyon, Peck, Tiger, Pine Grove, Bradshaw, Tintop, Humbug and Castle Creek Mining Districts. This volume is lavishly illustrated with rare photos and mining maps. **8.5" X 11", 218 ppgs. Retail Price: $19.99**

The Ajo Mining District of Pima County Arizona - This important publication on Arizona Mining has not been available for nearly seventy years. This volume contains rare insights into the geology and mineralization of the Ajo Mining District in Pima County, Arizona and in particular the famous New Cornelia Mine.   8.5" X 11", 126 ppgs. Retail Price: $11.99

Mining in the Santa Rita and Patagonia Mountains of Arizona - Originally published in 1915, this important publication on Arizona Mining has not been available for nearly a century. Included are rare insights into hundreds of gold, silver, copper and other mines in this famous Arizona mining area. Details include the locations, geology, history, production and other facts of the mines of this region. 8.5" X 11", 394 ppgs. Retail Price: $24.99

Montana Mining Books

A History of Butte Montana: The World's Greatest Mining Camp - First published in 1900 by H.C. Freeman, this important publication sheds a bright light on one of the most important mining areas in the history of The West. Together with his insights, as well as rare photographs of the periods, Harry Freeman describes Butte and its vicinity from its early beginnings, right up to its flush years when copper flowed from its mines like a river. At the time of publication, Butte, Montana was known worldwide as "The Richest Mining Spot On Earth" and produced not only vast amounts of copper, but also silver, gold and other metals from its mines. Freeman illustrates, with great detail, the most important mines in the vicinity of Butte, providing rare details on their owners, their history and most importantly, how the mines operated and how their treasures were extracted. Of particular interest are the dozens of rare photographs that depict mines such as the famous Anaconda, the Silver Bow, the Smoke House, Moose, Paulin, Buffalo, Little Minah, the Mountain Consolidated, West Greyrock, Cora, the Green Mountain, Diamond, Bell, Parnell, the Neversweat, Nipper, Original and many others.   8.5" X 11", 142 ppgs. Retail Price: $12.99

The Butte Mining District of Montana - This important publication on Montana Mining has not been available for over a century. Included are rare insights into the gold, copper and silver mines of Butte, Montana together with hard to find maps and photographs. Some of the topics include the early history of gold, silver and copper mining in the Butte area, insight into the geology of its mining areas, the local distribution of gold, silver and copper ores, as well their composition and how to identify them. Also included are detailed facts about the mines in the Butte Mining District, including the famous Anaconda Mine, Gagnon, Parrot, Blue Vein, Moscow, Poulin, Stella, Buffalo, Green Mountain, Wake Up Jim, the Diamond-Bell Group, Mountain Consolidated, East Greyrock, West Greyrock, Snowball, Corra, Speculator, Adirondack, Miners Union, the Jessie-Edith May Group, Otisco, Iduna, Colorado, Lizzie, Cambers, Anderson, Hesperus, Preferencia and dozens of others. 8.5" X 11", 298 ppgs. Retail Price: $24.99

Mines of the Helena Mining Region of Montana - This important publication on Montana Mining has not been available for over a century. Included are rare insights into the gold, copper and silver mines of the vicinity of Helena, Montana, including the Marysville Mining District, Elliston Mining District, Rimini Mining District, Helena Mining District, Clancy Mining District, Wickes Mining District, Boulder and Basin Mining Districts and the Elkhorn Mining District. Some of the topics include the early history of gold, silver and copper mining in the Helena area, insight into the geology of its mining areas, the local distribution of gold, silver and copper ores, as well their composition and how to identify them. Also included are detailed facts, history, geology and locations of over one hundred gold, silver and copper mines in the area . 8.5" X 11", 162 ppgs, Retail Price: $14.99

Mines and Geology of the Garnet Range of Montana - This important publication on Montana Mining has not been available for over a century. Included are rare insights into the gold, copper and silver mines of the vicinity of this important mining area of Montana. Some of the topics include the early history of gold, silver and copper mining in the Garnet Mountains, insight into the geology of its mining areas, the local distribution of gold, silver and copper ores, as well their composition and how to identify them. Also included are detailed facts, history, geology and locations of numerous gold, silver and copper mines in the area . 8.5" X 11", 100 ppgs, Retail Price: $11.99

Mines and Geology of the Philipsburg Quadrangle of Montana - This important publication on Montana Mining has not been available for over a century. Included are rare insights into the gold, copper and silver mines of the vicinity of this important mining area of Montana. Some of the topics include the early history of gold, silver and copper mining in the Philipsburg Quadrangle, insight into the geology of its mining areas, the local distribution of gold, silver and copper ores, as well their composition and how to identify them. Also included are detailed facts, history, geology and locations of over one hundred gold, silver and copper mines in the area 8.5" X 11", 290 ppgs, Retail Price: $24.99

Geology of the Marysville Mining District of Montana - Included are rare insights into the mining geology of the Marysville Mining District. Some of the topics include the early history of gold, silver and copper mining in the area, insight into the geology of its mining areas, the local distribution of gold, silver and copper ores, as well their composition and how to identify them. Also included are detailed facts, history, geology and locations of gold, silver and copper mines in the area 8.5" X 11", 198 ppgs, Retail Price: $19.99

The Geology and Mines of Northern Idaho and North Western Montana

See listing under Idaho.

Nevada Mining Books

The Bull Frog Mining District of Nevada - Unavailable since 1910, this publication was originally compiled by the United States Department of Interior. This volume also includes important insights into the geologic formations, faults and other aspects of economic geology in this Nevada mining district. Of particular interest are the fine details on many mines in the area, including their locations, histories, development and mineralization. Some of the mines featured include the National Bank Mine, Providence, Gibraltor, Tramps, Denver, Original Bullfrog, Gold Bar, Mayflower, Homestake-King and other mines and prospects. **8.5" X 11", 152 ppgs, Retail Price: $14.99**

Colorado Mining Books

Ores of The Leadville Mining District - Unavailable since 1926, this publication was originally compiled by the United States Department of Interior. This volume also includes important insights into the ores and mineralization of the Leadville Mining District in Colorado. Topics include historic ore prospecting methods, local geology, insights into ore veins and stockworks, the local trend and distribution of ore channels, reverse faults, shattered rock above replacement ore bodies, mineral enrichment in oxidized and sulphide zones and more. **8.5" X 11", 66 ppgs, Retail Price: $8.99**

Mining in Colorado - Unavailable since 1926, this publication was originally compiled by the United States Department of Interior. This volume also includes important insights into the mining history of Colorado from its early beginnings in the 1850's right up to the mid 1920's. Not only is Colorado's gold mining heritage included, but also its silver, copper, lead and zinc mining industry. Each mining area is treated separately, detailing the development of Colorado's mines on a county by county basis. **8.5" X 11", 284 ppgs, Retail Price: $19.99**

Gold Mining in Gilpin County Colorado - Unavailable since 1876, this publication was originally compiled by the Register Steam Printing House of Central City, Colorado. A rare glimpse at the gold mining history and early mines of Gilpin County, Colorado from their first discovery in the 1850's up to the "flush years" of the mid 1870's. Of particular interest is the history of the discovery of gold in Gilpin County and details about the men who made those first strikes. Special focus is given to the early gold mines and first mining districts of the area, many of which are not detailed in other books on Colorado's gold mining history. **8.5" X 11", 156 ppgs, Retail Price: $12.99**

Mining in the Gold Brick Mining District of Colorado - Important insights into the history of the Gold Brick Mining District, as well as its local geography and economic geology. Also included are the histories and locations of historic mines in this important Colorado Mining District, including the Cortland, Carter, Raymond, Gold Links, Sacramento, Bassick, Sandy Hook, Chronicle, Grand Prize, Chloride, Granite Mountain, Lucille, Gray Mountain, Hilltop, Maggie Mitchell, Silver Islet, Revenue, Roosevelt, Carbonate King and others. In addition to hardrock mining, are also included are details on gold placer mining in this portion of Colorado. **8.5" X 11", 140 ppgs, Retail Price: $12.99**

Washington Mining Books

The Republic Mining District of Washington - Unavailable since 1910, this important publication was originally published by the Washington Geologic Survey and has been unavailable for a century. Topics include the geology, rock formations and the formation of ore deposits in this important mining area of Washington State. Also included are hard to find details on the geology, history and locations of dozens of mines in the area. Some of the mines featured include the New Republic Mine, Ben Hur, Morning Glory, the South Republic Mine, Quilp, Surprise, Black Tail, Lone Pine, San Poil, Mountain Lion, Tom Thumb, Elcaliph and many others. **8.5" X 11", 94 ppgs, Retail Price: $10.99**

Wyoming Mining Books

Mining in the Laramie Basin of Wyoming - Unavailable since 1909, this publication was originally compiled by the United States Department of Interior. Also included are insights into the mineralization and other characteristics of this important mining region, especially in regards to coal, limestone, gypsum, bentonite clay, cement, sand, clay and copper. **8.5" X 11", 104 ppgs, Retail Price: $11.99**

More Mining Books

Prospecting and Developing A Small Mine - Topics covered include the classification of varying ores, how to take a proper ore sample, the proper reduction of ore samples, alluvial sampling, how to understand geology as it is applied to prospecting and mining, prospecting procedures, methods of ore treatment, the application of drilling and blasting in a small mine and other topics that the small scale miner will find of benefit. **8.5" X 11", 112 ppgs, Retail Price: $11.99**

Timbering For Small Underground Mines - Topics covered include the selection of caps and posts, the treatment of mine timbers, how to install mine timbers, repairing damaged timbers, use of drift supports, headboards, squeeze sets, ore chute construction, mine cribbing, square set timbering methods, the use of steel and concrete sets and other topics that the small underground miner will find of benefit. This volume also includes twenty eight illustrations depicting the proper construction of mine timbering and support systems that greatly enhance the practical usability of the information contained in this small book. **8.5" X 11", 88 ppgs. Retail Price: $10.99**

Timbering and Mining - A classic mining publication on Hard Rock Mining by W.H. Storms. Unavailable since 1909, this rare publication provides an in depth look at American methods of underground mine timbering and mining methods. Topics include the selection and preservation of mine timbers, drifting and drift sets, driving in running ground, structural steel in mine workings, timbering drifts in gravel mines, timbering methods for driving shafts, positioning drill holes in shafts, timbering stations at shafts, drainage, mining large ore bodies by means of open cuts or by the "Glory Hole" system, stoping out ore in flat or low lying veins, use of the "Caving System", stoping in swelling ground, how to stope out large ore bodies, Square Set timbering on the Comstock and its modifications by California miners, the construction of ore chutes, stoping ore bodies by use of the "Block System", how to work dangerous ground, information on the "Delprat System" of stoping without mine timbers, construction and use of headframes and much more. This volume provides a reference into not only practical methods of mining and timbering that may be employed in narrow vein mining by small miners today, but also rare insights into how mines were being worked at the turn of the 19th Century. **8.5" X 11", 288 ppgs. Retail Price: $24.99**

A Study of Ore Deposits For The Practical Miner - Mining historian Kerby Jackson introduces us to a classic mining publication on ore deposits by J.P. Wallace. First published in 1908, it has been unavailable for over a century. Included are important insights into the properties of minerals and their identification, on the occurrence and origin of gold, on gold alloys, insights into gold bearing sulfides such as pyrites and arsenopyrites, on gold bearing vanadium, gold and silver tellurides, lead and mercury tellurides, on silver ores, platinum and iridium, mercury ores, copper ores, lead ores, zinc ores, iron ores, chromium ores, manganese ores, nickel ores, tin ores, tungsten ores and others. Also included are facts regarding rock forming minerals, their composition and occurrences, on igneous, sedimentary, metamorphic and intrusive rocks, as well as how they are geologically disturbed by dikes, flows and faults, as well as the effects of these geologic actions and why they are important to the miner. Written specifically with the common miner and prospector in mind, the book will help to unlock the earth's hidden wealth for you and is written in a simple and concise language that anyone can understand. **8.5" X 11", 366 ppgs. Retail Price: $24.99**

Mine Drainage - Unavailable since 1896, this rare publication provides an in depth look at American methods of underground mine drainage and mining pump systems. This volume provides a reference into not only practical methods of mining drainage that may be employed in narrow vein mining by small miners today, but also rare insights into how mines were being worked at the turn of the 19th Century. **8.5" X 11", 218 ppgs. Retail Price: $24.99**

Fire Assaying Gold, Silver and Lead Ores - Unavailable since 1907, this important publication was originally published by the Mining and Scientific Press and was designed to introduce miners and prospectors of gold, silver and lead to the art of fire assaying. Topics include the fire assaying of ores and products containing gold, silver and lead; the sampling and preparation of ore for an assay; care of the assay office, assay furnaces; crucibles and scorifiers; assay balances; metallic ores; scorification assays; cupelling; parting' crucible assays, the roasting of ores and more. This classic provides a time honored method of assaying put forward in a clear, concise and easy to understand language that will make it a benefit to even beginners. **8.5" X 11", 96 ppgs. Retail Price: $11.99**

Methods of Mine Timbering - Originally published in 1896, this important publication on mining engineering has not been available for nearly a century. Included are rare insights into historical methods of timbering structural support that were used in underground metal mines during the California that still have a practical application for the small scale hardrock miner of today. **8.5" X 11", 94 ppgs. Retail Price: $10.99**